The Driving Force

The Driving Force

FOOD, EVOLUTION AND THE FUTURE

Michael Crawford and David Marsh

A Cornelia & Michael Bessie Book

HARPER & ROW, PUBLISHERS, New York
Grand Rapids, Philadelphia, St. Louis, San Francisco
London, Singapore, Sydney, Tokyo, Toronto

This book was originally published in 1989 in Great Britain by William Heinemann Ltd. It is here reprinted by arrangement with William Heinemann Ltd.

FIRST U.S. EDITION

LIBRARY OF CONGRESS CATALOG CARD NUMBER 88-45891

ISBN 0-06-039069-7

---

89 90 91 92 93 HC 10 9 8 7 6 5 4 3 2 1

To Sheilagh and Shandra

Contents

Acknowledgements

Much contemporary knowledge of biomedical systems comes from experiments on small laboratory rodents and observations on ourselves. Yet there is a wealth of different species, the study of which tells us much about the variation between species and the reason why we are different from rats and guinea-pigs. This was the philosophy which led Lord Zuckermann, as Secretary of the Zoological Society of London, to raise funds and create the Nuffield Institute of Comparative Medicine. It was this philosophy which attracted me to the Institute and so I owe a debt of gratitude to Zuckermann for making the research possible which provided much groundwork for this book. In the same context, I wish to thank the Institute's first Director, Dr Len Goodwin, whose broad vision and knowledge of biology was an essential stimulus and guide.

It was Africa which first shook my understanding of biology and man's place in it. As a chemist with an interest in the gap between medicine and chemistry, I worked and studied at the Royal Postgraduate Medical School to find out what it was all about. It was there that the first step to the comparative approach was insinuated by Sir John McMichael who persuaded my family and myself to set sail for Africa rather than Boston. 'The same people living no more than 100 miles apart have totally different disease patterns – one group has a form of heart disease quite unlike ours – why?'

The year was 1960 and the aim was to establish biochemistry teaching at the University Medical School at Makerere in Kampala, Uganda. Once established, I set about this task with a fervour for convincing my medical students of the importance of molecular biology and a condescending request to the late Dr Rex Dean to give them two weeks of lectures on nutrition. It took about three years to realise that this balance was hopelessly misplaced. But it was only on my return to the UK that the real difference between what we eat today and what man ate through his evolution struck home with enough force to awaken some dawning of realisation that nutrition mattered.

Makerere contained many people who were revelling in the wealth of opportunity presented by the comparative approach. Professor Shaper's early work made it clear that raised blood cholesterol and blood pressure levels were not a feature of *Homo sapiens* but only of Western peoples. Dennis Burkitt's studies on lymphoma made their contribution to Africa and science and he too, on his return to the UK, suddenly realised the striking difference in the nature of the stools people passed, how that related to their food and to the high incidence of diverticular disease and colon cancer in the UK which was unheard of amongst our African colleagues. Professor Jack Davies, through his contacts at the NIH of the USA, stimulated much of the comparative disease approach by raising the necessary, and for those days copious, funds. Kris Somers and he pioneered much of the descriptive work on endomyocardial fibrosis – the heart disease of the Africans but not the Europeans. Encouraged by Jack, Peter Turner and Brian McKinney, some of us felt bold enough to suggest there might be a nutritional reason.

In parallel to the human research there was also a flourishing group under Dr Richard Laws who was head of the Nuffield Unit for Tropical Animal Ecology. The opportunities for wildlife research were truly remarkable. Dick's studies on elephant populations provided science with a witness of the principles of catastrophic mass-extinction happening in our time. It was a brilliant stroke to incorporate Dr Sylvia Sikes with her interest in studying the pathology associated with this event. In amongst all this, Professor David Allbrook escaped from anatomy teaching to delve into the paleontological past of Africa, and gave me my first lesson on the subject.

My research was greatly assisted by Inge Berg-Hansen who ran the biochemical research laboratory ably supported by Lawrence Mwasi and Keffer Nguli. Neil Casperd from pharmacology was interested in developing the use of drugs via a crossbow, to anaesthetise and translocate wild animals. The common interest in wildlife led to remarkable friendship and collaborative programme on the difference in the nutrient value of wild species compared with the end product of modern farming. In this respect we were given much help by Mr Silvester Ruhweza of the Uganda Game Department. Phyl Msuya was a constant companion on safari and in the machinations to create the Dar es Salaam Medical School where he later became head of biochemistry and then assistant Dean. Several of our sample analyses were carried out in his department by Aaron Munhambo.

The post at the Nuffield Institute in London offered new analytical technology to apply to the interesting problems which arose out of comparing Africa and Britain. Man is biologically unlike laboratory or domestic species and remains unselected for any particular purpose. He is physiologically still a wild animal but the foods he eats today have only been with him for a short period of a few centuries or less. By then technology could define at least some of the relevant differences and it did: it would simply have been impossible for man throughout his evolution to have eaten such huge amounts of fat (and particularly saturated fat) during the 5 million year period when he relied on wild foods.

My understanding of the biochemical mechanisms was fired by a lecture by Professor Hugh Sinclair in 1966 at a Zoological Society Symposium which explained what fats meant to atherosclerosis and heart disease. It then became clear that animals including ourselves did not only need protein to build our bodies but that they also required fats: the problem was that they had to be the right kinds of fat. I am convinced that if Hugh gave that same lecture today, there is little he would need to change: he was far in advance of the field.

Of course fats were not the only difference between the foods on man's natural background and modern products; but it was clear that they were pretty important to the epidemic of atherosclerosis and heart disease. Mary Gale, who was working with me at the time, made a simple suggestion which actually became quite important.

She commented that everyone was working on atherosclerosis so if we were interested in fats, why not work on the brain which is made more of fat than anything else! The establishment of the day did not like the idea that our cows might be killing us (as the *New Scientist* put it). The antagonism from the establishment helped to cement the change in direction for which they are now to be thanked. The brain after all was the most signficant feature that made *Homo sapiens* different from chimps and buffaloes and very little was known about its chemistry and even less about its links with nutrition. It was tacitly assumed that the brain could make all the fats it needed but that idea was based on poor analytical chemistry which failed to reveal the presence of long-chain essential fatty acids.

In the investigation that followed, there were many who contributed far more than I; Andrew Sinclair joined me from Canada and soon found the level at which you could feed enough essential fatty acids to let laboratory rats become pregnant but not enough to satisfy the requirements for proper brain development. Whilst everyone else was working on rats, which can perform metabolic miracles with nutrients, John Rivers contributed substantially by showing just how difficult it was for the cat to make the long-chain essential fats which it needed for its brain: the cat relied on other animals to do the job for it. John was also a most stimulating colleague: he could always find something new to say. It was for this reason that David Marsh and I were glad that he read and commented on this text.

In the meantime, comparisons of different species was showing that brain size seemed to be related to amounts of these special essential fatty acids that the lifestyle of a species could achieve. Putting the experimental and the comparative data together spelt out a long-range link between nutrition and the brain. Dr Ahmed Hassam then filled in many of the details and Professor Pierre Budowski descended on us from Israel to forge a new dimension in our understanding of the need for both families of essential fatty acids for the brain. In the meantime Wendy Doyle tested the relevance of the data to pregnant mothers. Whilst Patrick Drury built all the necessary computer software, Martin Leighfield, Ann Lennon and many others conducted the difficult analyatical procedures. The medical aspects could not have been envisaged without the active participation of Drs Bernard Laurence, Kate Costeloe and Alison Leaf and many others.

The applied aspect of our studies was greatly enhanced both scientifically and from being given a sense of purpose through the support from and close co-operation with Action Research into Multiple Sclerosis both from the London and Northern Ireland groups. John Simkins quickly took on board the significance of nutrients to the brain in relation to MS, and through their support, and the encouragement of Ann Walker's group in Belfast, much new ground has been covered and questions raised, the answers to which we believe can lead to beneficial management of the disease.

Many others in the small band of international workers also made special contributions to the knowledge used in this book. There are too many to mention but Claudio Galli from Milan developed, at the earliest stage, a similar interest and helped present the case for essential fatty acids at an expert meeting called conjointly by FAO and WHO in 1977. Ralph Holman, Bob Ackman, Jim Willis, Bill Lands, Serge Renaud, Michel Lagarde, Peter Ramwell, Howard Sprecher and a few more of us banded together to set up two international congresses for the field, the first as a Golden Jubilee celebration for the discovery of essential fatty acids and prostaglandins and the second to mark the Nobel Prize awarded to Bergstrom, Sammuelson and Vane. The publications of the proceedings were landmarks as well as being collections of vital, readily accessible information on the subject.

Sir Alister Hardy's theory of the aquatic ape was brought to my attention by Stephen Cunnane who had himself written on the subject. Previously, Andrew Sinclair and I had dreamt up the idea that the difference between the carnivore and the herbivore was based on the contrasting availability of the fatty acids specifically required by the brain. This idea was presented at a meeting on the brain organised jointly by the CIBA and the Nestlé Foundations in 1972. The biochemistry regarding the big human brain did not fit with the savannahs which everyone saw as the site of origin for *Homo sapiens*. How could one challenge this view which was held with such strength, simply on some biochemical calculations which suggested a role for seafoods? When Stephen pointed out that the physiology fitted with a marine origin, then 'Homo aquaticus' was born. There will doubtless be many who will argue against the idea of 'Homo aquaticus', just as they dismissed Hardy's original hypothesis: however, the agreement between the physiology and the biochemistry in support of an aquatic origin presents a formidable case. I have to thank Dr Stewart Barlow as Director of the International Fish Meal Manufacturer's Association for offering the opportunity to present the biochemical case at their Munich AGM in 1985. He had lost a key speaker so asked me to fill the gap and tell them how important fish was in the diet! More scientific detail was added at the Membrane meeting organised by Dr Alexander Leaf and his colleagues in Crans sur Sierre two years later and led to Chapter 9, the request by the Swedish Society of Medicine and the Swedish Nutrition Foundation to review the evidence on diet, heart disease and cancer provided the basis for Chapters 12 and 13. David and I are particularly grateful to Maurice Temple-Smith for his patient and valuable assistance in editing the manuscript.

Lastly, I need to mention Bill Pirie who also read the text. This does not mean to say he agrees with anything we say. The comment I want to make is that much of my own insight owes its origins to listening to him and reading what he had to say about biology. It was he who, for example, pointed out to me that the squirrel had the same relative brain size as *Homo sapiens*, a fact which assumes some importance in the discussion on the origin of the 'big brain'.

Science is moving too fast for any single person to keep pace with the rate at which knowledge is accessed. For this reason, the arguments presented in this book are offered from the limited domain of biochemistry in which we have some expertise. The animal examples are again confined to but a narrow aspect of the products of evolution for the same reason. The hope is that their anaylsis may assist in building a wider view of the events which shaped species and ultimately led to man. It is with humility that David and I have set down the words to present a new view of evolution. It is basically a chemical view but we hope is complementary to conventional theory.

M. A. Crawford, London, 1989

Thanks to all the doctors and scientists who patiently explained their theories during my research. To all in the Library at the Institute of Zoology, and in the department of Biochemistry and Nutrition at the Nuffield Institute of Comparative Medicine for their help over seven years; and most particularly to my wife Chandra for years of patience and understanding, for reading, rereading, and reordering innumerable versions of the text, and especially for achieving the final edit of the history of the evolution argument. Without her, the book in its present form would not have been achieved.

D. E. L. Marsh, London, 1989

The Driving Force

Introduction: the order of life

The multitude of living forms with which we share our rich and beautiful world strikes us with its amazing and incredible diversity. From the moss on the rock to the great oak tree growing above it, from the sea-squirt to the whale, from the single-cell bacterium to man himself, the span of life is immense. How could it ever come about that the barren minerals of any empty planet, themselves typified by the monotony of the moon, somehow shaped themselves into living beings, which changed and multiplied to produce the great spectrum of life that we see all around us?

Men have made many attempts to solve this mystery of life. From the earliest times, philosophies and religions have proposed their answers. But in the last century a single theory has come to be generally held, the theory of an unfolding evolution. In particular, scientists and public alike now accept what they take Charles Darwin to have said about 'random variation' and 'the survival of the fittest'. The force that drove forward the progress of life was simply small, random changes in genetics leading to the development of forms that were fitter or better equipped to survive in the prevailing conditions. Existing life forms would from time to time happen to give birth to progeny which differed in some way from their parents. These mutants would generally be ill-fitted to their environment and would die, but occasionally some would be better fitted for it, and these would survive. In time, as the struggle for existence went on, they would displace the older and less well adapted form. So plants and animals slowly changed until, after a long accumulation of mutations, each one perhaps insignificant in itself, a new species would emerge. The idea of natural selection was thus conceived.

This theory has survived many criticisms and undoubtedly it is one of the greatest intellectual leaps of mankind. It has taken on board the science of genetics and, in recent years, the understanding of cells and their reproductive mechanisms. Out of these various theories it has formed a synthesis that seems impregnable. Yet it has not been without critics, and criticism has not come only from people affronted because the doctrine of evolution seems to contradict the sacred writings.

The thesis however contains one important oversight. Darwin considers how species changed little by little to become what they are today. He does not discuss how it all happened in the first place.

There is also a separate problem which is not of Darwin's making. He rewrote his thesis several times and attempted to find the solution to a problem which worried him. How, exactly, did species interact with their environment? By asking that question he recognised that natural selection was not the only operational force in evolution.

Darwin said there were two factors of paramount importance in evolution: natural selection, and what he termed 'conditions of existence' – we would say chemistry, or the environment. He argued that as natural selection was dependent on 'conditions', then conditions were the most important of the two controlling factors.

So original Darwinism perceived the environment to be a major directive force, although a precise mechanism of how environment interrelated with natural selection was not clear. It was this problem that Darwin spent the rest of his life trying to work out.

People who came after Darwin rejected this question and neo-Darwinism now sees selection as the only mechanism for evolutionary change. Through briefly reviewing the history of evolution theory, it can be seen, firstly, how this came about, and secondly, how a contemporary understanding of biochemistry can throw light on the sort of mechanism for which Darwin and others after him were looking.

There are several recurring objections to the narrower 'post-Darwin' view of evolution. One objection can be put in the form of a simple question: 'Has there been enough time?'

Many scientists have questioned the 'collection of small, random changes' as a mechanism for the origin of life, let alone new species. Indeed, Albert Szent-Györgyi, the scientist and Nobel prize winner, has presented the case that the probability of life emerging by chance is *zero* (Szent-Györgyi, 1977). The Cambridge astronomer, Sir Fred Hoyle, like Szent-Györgyi, was so convinced by the mathematical implausibility that he abandoned the idea of evolution taking place on our own planet and proclaimed that it must have arrived from outer space. In this book, we shall simply suggest that the principles of chemistry and nutrition remove the element of chance.

It is instructive to look at a well-known analogy that was used to justify the chance process suggested by Darwin. The analogy is commonly attributed to T. H. Huxley, a leading figure in British science, who is supposed to have invited his readers to imagine a group of monkeys, each at his own typewriter, hammering away at the keys in a totally random way. In time, it is said, they would type out the complete works of Shakespeare. Mathematically this is perfectly true. In time they would; but the question is, how much time? To test this out, we wrote a computer programme to calculate how long it would take for one of Huxley's monkeys to type out just the first line of a single Shakespeare sonnet. The programme ran, and the computer informed us that the answer was too large a number for it to print. At this stage we realised that you do not need a computer, simply a table of logarithms. By converting the problem to logarithms we were able to get a result, and it was quite startling. If a monkey went on untiringly hitting three or four keys every second – say two hundred a minute – the job would take just under thirty billion years. If he had started typing at the moment when the universe began, he would still be nowhere near halfway to producing that first line of verse.

What we forgot was that a typewriter has at least 45 keys not 26! (And that does not include upper and lower case.) When we told N. W. Pirie about our problems with the computer he remarked that when he was at school, a master tried to use the same argument about monkeys. He immediately pointed out that with 45 keys and (say) 30 letters in a line, guessing that the log of 45 was 1.7, you had a 50 figure number for the possible permutations. A computer is not needed.

This, of course, no more proves Darwin wrong than Huxley's analogy could have proved him right. What it does show is that arguments about what will occur by random change are valueless if we do not consider the actual logistics compatible with the mechanism. Although science has yet to describe the full range of numbers and the chemical mechanisms involved in evolution, it is astonishing that so many have swallowed this monkey business!

Hoyle is by no means the only critic to believe that, for a simple model of the evolution of life and natural selection, the numbers do not fit. Professor C. H. Waddington, a geneticist who did much experimental work with animals, has compared the theory of evolution through chance to a builder 'throwing bricks together in heaps' in the hope that they would 'arrange themselves into a habitable house'.

These simple demonstrations give some idea of the sizeable mathematical constraints which convinced Szent-Györgyi, Hoyle and Waddington, among others, that evolution could not have happened by chance alone.

So how could it have worked? Chemical molecules are themselves built to specifications which depend on the nature of the chemicals reacting together. This property of chemicals is not lost simply because the chemical is part of a living system. Hence the chemicals which interacted to produce what we now call the genetic code and the further interaction with proteins, RNA and other molecules, had to be arranged in response to the options set by the laws of chemistry and physics. This targeting by the laws governing how one chemical reacts with another simply takes the guesswork out of the game.

However, those with faith in the lottery have fought back. A computer programme, if given rules, can readily assemble meaningful phrases out a jumble. Richard Dawkins uses this technique in his book *The Blind Watchmaker* (Dawkins, 1986). To solve the Huxley paradox he analyses the data on the basis of cumulative selection whereby the first selection restricts the possibilities for subsequent selections; in so doing the number of options is dramatically reduced to a point where it is all possible by chance. However, this technique does not solve the paradox. What Hoyle and Szent-Györgyi are worried about in the first place is not selection but the origin of life itself; selection comes at a much later stage. The difference between one species and the next is tiny in comparison with the difference between the presence and absence of life. If, however, the beginning of life was the dedicated work of chemistry, then we need not consider chance.

The rules of chemistry are not quite blind and were made long before life evolved. It is not a question of a blind maker but one with its eyes wide open to the laws of chemistry. The important distinction between blindness and rules is that rules allow predictions to be made about associations and directions. This is the basis of the scientific method and when scientists get it right, the predictions are proved correct. Rules enable us to predict what happens when A meets B. Hence chance can be replaced by an inevitable chemistry and the Huxley numbers are brought down to a manageable size.

The numbers involved in the creation of life by random events are difficult to comprehend and it is also difficult to ascribe to chance the simpler question of the evolution of the sophisticated hierarchy of living things. This difficulty is illustrated by a consideration of what needs to happen for what appears to be a single new modification to succeed.

Let us consider the long neck of the giraffe which has been used often in the debate on natural selection. Those animals with slightly longer necks than their friends survived because they could reach higher into the trees and so gain the advantage of reaching foods that others could not.

If, say, some forerunner of the giraffe was born with a slightly longer neck than its parent, the greater length would be valueless if the animal did not also modify its heart to pump enough blood at increasing pressure to meet the greater demand, and its vascular system to take the strain. Its nervous system too, must change. Unless it develops some pressure-regulating valves in the blood vessels leading to the brain, then every time it lifts its head from drinking water, it will simply black out as the blood drains from the brain. Its pattern of behaviour must also change if it is to seek its food from the tops of trees. If the food that is now brought within its reach is different from that which its shorter-necked parents browsed on (and we shall see that it is), then its stomach and digestive system will also have to be modified. So it is not just the time needed to develop the gene for a long neck; the time taken for all these other essential changes to coincide also needs to be taken into account.

Unless there is some mechanism linking these changes, the chances against them all happening simultaneously are enormous. The changes taking place are not just cumulative but are also co-ordinated. Indeed, throughout the different lines of evolution, they are so co-ordinated that it is far from likely that chance alone could have produced the great variety of living forms that have come into existence, even in the two to three billion years that have passed since life first began. Yet the creation of the animal phyla with all their great diversity and subsequent change to the life forms we know today, occurred in less than 600 million years.

Cumulative selection could have provided the answer to the evolution of the giraffe's neck but would it have done so by chance? If you say that the rules for cumulative selection are laid down only by chance then you have to ask, 'What is the chance that this or that restriction is imposed?' Then you are back where you started. This is not say that cumulative selection does not occur, but if you want to extract yourself from chance, then you have to say that at some stage chance is replaced by a rule or by order; a physical or chemical mechanism is required. The important question is: who makes the rules for the success of the selection and what are they?

There have been other objections, too, and we shall be looking at them in later chapters. If change did indeed occur randomly through the accumulation of many tiny modifications, then it would be likely to occur at a steady pace. One of the most important objections to this theory concerns the many examples of sudden branching or new development in evolution, referred to as the *discontinuities*. Conventional theory predicts a steady drift of change. However, for the most part change has been far from steady. Immensely long periods of stability have been punctuated by the sudden emergence of a range of new species, with no fossil record surviving of any intermediate forms. For millions and millions of years there is no change, then suddenly the old collapses and the new appears. S. J. Gould drew attention to these discontinuities as recurrent themes punctuating evolution which are difficult to explain through chance (Eldredge & Gould, 1977).

There has to be some reason why the evolutionary forces for real change did nothing for so long, then leapt into top gear before suddenly coming to a stop. If random mutation was continuous, as the accepted thesis supposes, was its potential for innovation held at bay? If so what held it at bay and, even more interestingly, what was it that suddenly released the mechanism for change? Indeed, what called the sudden excitement to a halt?

For example, the dinosaurs ruled the earth for 160 million years, and then over a million years they disappeared and quite a new family of animals, the mammals, took their place. Flies and bees emerge in the fossil record as if from nowhere. There are no intermediate forms. Some genetic changes have occurred with time but their form has been largely unchanged. Much the same is true for the modern mammals.

To explain these discontinuities Hoyle has resorted to visits from outer space. Others have suggested recurring catastrophes – the whole world darkened with the dust of enormous volcanic eruptions, or struck by some giant meteor or asteroid. This book will suggest a less dramatic agency of change.

Another problem lies in the basic chemistry of all the life forms we know about; if changes had been truly random we would expect them to have affected this biochemical level of organisation as well as others, but in fact they have not. The way in which modern plants absorb sunlight, use its energy to promote chemical reactions, and excrete oxygen into their surroundings is the same (as far as we know) as that of the earliest forms of life. Still more curiously, we shall see that the chemistry that transports oxygen in our human bloodstream is the same in all mammalian species. Indeed, the key processing systems in our bodies have not radically changed from those of the first microscopic creatures that drifted in the primeval oceans. In a world of chance events such persistence over millions and even billions of years needs explaining.

This uniformity of structure and function may be because a wide range of different lines of evolution converged on the same answer as to what is best fitted for the purpose. On the other hand, the reason for uniformity may simply be, as N. W. Pirie remarked in 1972, 'that in similar environments, only one form of biochemistry is possible'. He argues that, whilst evolution comes from selection, 'it is not always certain from which direction the selection comes. Chemical structures first appear and then uses or new uses are made of them later.' Indeed, he believes that ideas on evolutionary progress are basically an aesthetic appreciation of morphology, 'coloured by anthropomorphism'. For example, we are so familiar with the hand and its function that it is difficult to imagine any better design existing either in the past or in the future. The design of the hand seems to be so perfect that it becomes a perfect piece of evidence for natural selection.

But it is not only the *chemistry* of life that resists change. If we compare the dolphin's flipper and the arm and hand of a man we see that the basic plan of the bones is identical, and this is odd. If an engineer had to design two 'perfect' pieces of apparatus, one to propel a boat and the other to manipulate small objects, it is hardly likely that he would use the same structure of tiny rods for both mechanisms, simply varying their lengths and thicknesses. Yet that is what has happened with men and dolphins, and with many other species as well. The changes to their forelimbs may have been selective, but they have not been random. The sizes of the individual bones have varied greatly but, curiously, their number and relative positions have not changed at all.

This example suggests that there is some not-yet-understood restraint that makes certain kinds of change less or more likely to occur than others. Alternatively, there is some biological mechanism which uses the same ground-plan and simply makes more or less of this or that. Did the dolphin evolve the design of the hand inside its flipper? Or did the arm bones shrink into the body and surplus skin join the fingertips to make flippers? The design of the hand as evidence for natural selection is now perhaps less convincing. We can ask: 'Is it more likely that the dolphin arrived at that shape and form of the arm and hand from a different and unknown starting point, or did it start from a similar plan to the human and was subsequently modified?' To us the former seems unlikely. And if that did not happen then evolution can be shown to work through changing shape and size but while retaining the basic plan.

As in these details so in the broadest view, evolution displays certain curious streaks of conservatism on the one hand and an astonishing diversity on the other. The most fundamental grouping of living things that biologists recognise is the phylum. There are about 30 phyla although the exact number is still a matter of dispute. What defines a phylum is that its members share a certain overall plan or strategy of life. There are, for example, single-cell systems such as algae and protozoa, and multicellular systems such as trees and animals with backbones. Plants get their food from air, sun and soil, animals by chasing it. Fungi, on the other hand, feed on organic matter as animals do, but stay rooted like plants. Again, there are different ways in which living systems are organised. The design that we humans have followed is based on the principle of two symmetrical halves – two eyes, two kidneys, two halves of our brain, each limb matching another on the opposite side. A starfish or a limpet is not like that: its design is radial not bilateral like ours, whilst an oyster is different again.

Evolution, it seems, gets itself into certain grooves from which it finds it hard to escape. Again, that is not random. Many species, successful for long periods, have died out altogether. Natural selection presumably worked well for them for these periods, but ultimately the changes needed to adapt them to a changing environment were simply not within the repertoire available to them. In fact, classes of animals get stuck in grooves. The elephant, whatever the pressure on it, could never randomly change into a carnivore, nor the lion learn to live on grass. If by random changes we mean that a change of any kind is as likely as a change of any other kind, then clearly this is not how evolution works.

Although the rules of chemistry can be thought of as absolute, they do allow for the possibility of competition and we might envisage the element of chance taking the organism in one of a number of directions. The chemistry of the large molecules is complex and the number of letters in the genetic code so large that even obeying the rules of chemistry, the number of possible permutations is such that room for random variation would be expected alongside certain fundamental dictates. One example is the favoured theory on immune system function, proposed by A. L. Burnett, which is decidedly dependent on chance. Lymphocytes are generated with a vast repertoire of antibody variation. When a cell matches an invading antigen, it is stimulated to divide and so produces a clone of identical cells making an identical antibody. Further stimulation by the same antigen multiplies the rate of reproduction leading to a rapid expansion of the defence system. This mechanism is an elegant example of natural selection operating within our own bodies which would rely specifically on the chance production of antibodies in the hope that one will provide a chemical match for the offensive agent.

A better known example of natural selection occurred in the soot-laden atmosphere and on the bark of trees in the days of the coal-burning industrial belt of England: dark coloured moths, successfully camouflaged against the pollution-blackened trees, survived, whilst the lighter coloured moths were eaten by birds providing a visible example of Darwinian predation at work.

Whilst we accept that natural selection is a factor, we believe, as Darwin also did, that it was not the only mechanism in evolution and that chemistry operating through the environment is at least as important.

The fundamental rule of chemistry is that those options which are thermodynamically most stable are the ones that persist. Chemistry will find a consistent direction, just as water runs downhill. A rock is hard and stable on the surface of our planet only because the temperature makes it so. The fact that the molecules in our bodies are more complicated than those in a rock does not mean that the laws of chemistry and physics are suspended.

If evolutionary change *was* obeying simple chemical laws, the persistence of traits, the problem of the evolutionary time-scales and Gould's discontinuities would all be explained. Instead of wasting time exploring innumerable dead ends, life would move forward according to a programme dictated by the nature of chemistry and the environment. The presence or absence of chemicals like oxygen, or nutrients like vitamin A, make certain directions possible or impossible. When chemical support systems for a particular line of species are exhausted, the biochemical system cannot cope, the line collapses and creates a new discontinuity.

In a sense, the example of Burnett's theory of immune system function is not only an example of specific selection but also an example of the interaction of external influences. Without the specific antigen matching an existing structure on the cell surface, that clone or race of cells would not become dominant.

The cells of the immune system are very different from the fertile egg and this process of differentiation may offer insights into evolution. How does the combination of two single cells, the sperm and ovum, turn into an eye or a hand? This important question has been tackled in a book published in 1987 by Norman Maclean and Brian K. Hall, *Cell Commitment and Differentiation*. They start with our present understanding of cell differentiation which offers two alternatives. Cells become skin or muscle because certain genes are (i) deleted or (ii) suppressed. Almost universally the latter strategy has been adopted so that the cell nucleus from any part of the body contains all the original DNA code and can be reused to produce a new animal. They comment that the recent rapid advance in knowledge of gene sequences might 'beguile' one into thinking that this process of differentiation is well understood. 'That this is not so is simply because knowing the sequence of a particular gene, or even how it is regulated, does not itself provide an insight into how a cell becomes committed to a particular fate.'

The remarkable similarity between different stages of embryonic development and the history of vertebrates, has raised the thought in the minds of many that embryonic development and cell differentiation emulate evolution. At about four weeks of age it is difficult to distinguish the human embryo from any other mammal; it has a long tail, a single-chambered heart like a fish, and a circulation with loops to four gill-arches, also like a fish. It seems as though it first organises itself in a manner similar to the fish and then develops this into the plan for a human.

There is indeed an interesting similarity between cell differentiation and evolution. Cells differentiate and become committed. Animals evolved and became committed. The question that cell differentiation and evolution both ask is: what is it that directs the paths which they follow? The question which cell commitment and the rigidity of evolutionary lines also ask is: what holds them in place? i.e. Why is a limb a limb and why does it stay as a limb? Why is a cow a cow and why does it not turn into a lion? is basically the same question. Add to this the rigidity of genetic codes for certain key systems (e.g. oxidative enzymes) and the question becomes: was evolution cell differentiation on a grand scale?

As Maclean and Hall point out, other than knowledge of anatomy, surprisingly little is known about cell commitment and differentiation. Similarly, little is known about regulation of shape and size. Although codes in the DNA may provide for 'amounts', we need to know about the chemical reasons for change: why does this chemical increase and that one decrease in amount? If such matters can be controlled internally by chemical signals, can external chemicals also act as messengers?

Most gardeners will have used plant rooting compounds to encourage faster rooting of cuttings. These plant auxins are in fact messengers applied externally which stimulate DNA synthesis and so increase cell division and the elongation of the cell through making more cell wall materials. The effect is to encourage root formation where before there were apparently no root cells. The plant rooting hormone is indolylacetic acid and it also co-ordinates with gibberellic acid to act on undifferentiated plant cells. In plants which are capable of producing male and female flowers, a high proportion of indolylacetic acid will promote female, whilst more gibberellins result in male, flowers.

Tadpoles will metamorphose, lose their tails and develop legs when stimulated with thyroid hormone. If production of the hormone is prevented, they do not develop into frogs. The thyroid hormone is a simple ring structure which contains iodine. The obvious conclusion is that if there were no iodine on the planet there would be no frogs.

The World Health Organization's map of the incidence of goitre shows that it is found mainly in mountainous regions. These are areas where the soil has had its trace elements washed away by rain and melting snow. A low iodine content leads to inadequate thyroxine production. In the human this in turn leads to a change in shape which is most easily recognised in the neck, face and eyes. In situations where the iodine content of the soil and food is very low, the change results in impoverished brain development.

These simple examples suggest that the common view that 'food is just food, but is there enough of it?' is inadequate. There is mounting evidence that it is not just the *amount* of food but its *qualitative composition* which matters. Indeed the contemporary debate on diet and its relation to Western heart disease and cancer is based on exactly that premise.

The question we are asking in this book is whether or not nutrition was a determinant in evolution which operated on – even directed – the basic plans of living systems. There is by no means enough evidence to spell out all the details but such evidence as exists is compelling. It suggests that just as the beginning of life was determined by chemistry so its later forms and expressions were also chemically shaped: not exclusively – sometimes chemistry would operate independently, sometimes in tandem with selection, and sometimes selection would operate alone.

We are suggesting that the origin of the planet and, subsequently, the origin of life was determined by chemistry and physics operating within the conditions which prevailed at the time. Once the conditions of temperature and pressure had stabilised, life remained relatively stable. We shall argue that there were crucially important changes in the conditions which brought about biological change.

Because the time-scales involved in the life of the planet are so huge, it tends to be taken for granted that the conditions on the planet are stable. This is far from the truth: they have been and are changing constantly. Even the rain since the beginning of the terrestrial environment has been washing salts and trace elements and waste products from the land into the sea and changing both. If the North Sea and the Baltic can be affected in a few decades by fertilisers, think of the effect of rain and climate on land over million year time-spans.

Although there may have been some form of stability in the physical conditions of the planet, chemical change nonetheless continued – in part as a result of the living systems, and partly through inexorable changes in the redistribution of the elements on the planet. At first these changes were in simple chemicals coalescing into living systems, but as life became more complex, chemicals became food. We will show how changes in the chemistry of food and differences in choice of foods are associated with different evolutionary paths. The implication is that change in food or choice of food is an evolutionary instrument.

For one-and-a-half billion years or more, the blue-green algae dominated life on the planet. They produced oxygen, and what was ultimately a destructive pollutant to the algae was both a mutagen and essential nourishment for the animals which were to come. The evolution of these air-breathing systems simply had to wait for the advent of oxygen. When it came it added the missing element to what was otherwise a rich niche for animal life. The coincident appearance of animal life is too much to ask of chance. The logic which this example suggests can be applied to us today in a manner that is important for our future.

While arguing this case in the context of the history of life on earth, and the successive waves of evolution, our main aim is to argue the case that nutrition has played a major role in shaping human history. This discussion leads naturally to the very strong probability that the recent changes in our own food have led to degenerative disease. We shall argue that the contemporary nutrition-related diseases which are specific to the different nutritional patterns of different countries are an example of this aspect of evolution in progress.

The report of the Surgeon General on health in the USA, 1988, should remove any remaining doubt about the relationship between diet and heart disease and the need to change diets to prevent this and other nutrition-related diseases. This is another way of saying that past changes in food have brought about diseases which now affect the populations of certain parts of the world. The degenerative diseases of the nervous and vascular systems which have become so common are basically changes in form. In one sense, they are the opposite of the co-ordinate changes of the development of the heart and the blood vessels in the giraffe.

The extension of the above argument is that changes in chemistry or food can influence the expression of the biological system. This means that the food we or the lion ate throughout our separate evolution had some bearing on what we and the lion are today. Nutrition will also play an important role in deciding our future.

If we understood enough about chemistry, we should know, for example, why the first self-replicating molecules and then the first living things took the forms they did, and why, in time, they gave way to certain particular successors, from which all later forms of life descend. It was not a matter of chance: it was cause and effect. Given the nature of the environment and its chemistry, there was only a limited number of paths which chemistry, and hence evolution, could have followed. Perhaps there was only one.

1 The great debate

In the great debate on how all the numerous forms of life arrived in the world there were two different controversies, essentially separate although often confused: (i) creation versus evolution; and (ii) natural selection of the fittest through competition and the struggle for survival versus an environmental stimulation. Did all the various species come into existence just as they are today, or did they evolve from earlier and simpler forms? If so, how? Today, no one who is willing to accept hard scientific evidence holds the creationist view, and even the Christian church, which at first vigorously opposed all evolutionary ideas, came to accept that the truth of its message does not depend on the literal truth of God bringing all living creatures into being, fully formed, in the six days allotted by the Book of Genesis. Darwin is held to have settled that dispute for good.

Regarding the second question, environmental stimulation versus natural selection, one can wonder to what extent the impulse for change comes from within the organism itself, and how much from its environment. Can the environment by some means induce differences in a plant or animal, independently of selection, and can such change be reproduced in its offspring? Or is the environment simply a filter eliminating through natural selection most mutations, and only allowing the fittest to survive? In either case, what is the actual mechanism of change? What is the origin of diversity? What indeed is the origin of life . . . the real origin of species, and what does the answer to that question tell us about our own origins? Darwin's views on external conditions versus selection were ambiguous. They were never fully clarified by Darwin himself, nor were they quite what his followers made them out to be.

It is with these questions that this book is concerned, and it is interesting to see how far back in human thought their origins go. All the great religions have had their explanations of how the world came into being. The Hindu faith saw Brahma as the source of everything: 'as sparks come out of the fire, so the whole universe comes out of Brahma'. Buddhism sees all of life as cause and effect in symbiotic interaction. The Yogic philosophies saw causes as 'subtle manifestations' and effects as 'gross manifestations'.

It was the Greeks who initially posed questions in something like the form in which they are posed today. Aristotle was one of those who saw life forms developing according to their own inner logic. For him there was a scale in nature, a progressive sequence of complexity culminating in man. The most primitive organisms arose by spontaneous generation, but some internal principle impelled them toward greater and greater perfection, so that a complete gradation existed between minerals and plants, plants and animals, animals and man.

In the 5th century BC Empedocles was the first to argue the possibility of the origin of the best-fitted forms arising by chance rather than through any sort of mystical design; Heraclitus introduced the concept of conflict and struggle for survival, and Democritus, one of the best known of the Atomists, believed in adaptations of individual structures and organs. But most evolution theories up to Darwin's time were theistic.

Thoughout the Christian era the debate was suspended. In the West, once Christianity became orthodox and allied to the State, the Church taught that the question had been settled by the Bible, and for the next 1500 years it had the power to enforce that doctrine, and to exercise controls to maintain it. The widespread and passionate dislike of such controls could eventually have assisted the rejection not only of religion, but also the theory of human origin which went with its philosophies: i.e. Christianity and creationism. Charles Darwin's own view of the Christian religion was probably shared by many others: 'I can indeed hardly see how anyone ought to wish Christianity to be true; for if so, the plain language of the text seems to show that the men who do not believe, and this would include my Father, Brother and almost all my best friends will be everlastingly punished. And this is a damnable doctrine (Darwin, 1876).

The first alternatives to creationism were published in several different European countries in the 19th century, and it is not surprising that they tended to have a Greek flavour. Two of the most important of these early evolutionists, the Englishman Erasmus Darwin (Charles Darwin's grandfather) and the Frenchman

Jean-Baptiste Lamarck, can be seen as refining Aristotle's ideas. Lamarck's thesis was of an organically unfolding evolution. His theories were theistic and teleological. He evoked a 'life force' in all living creatures, and a rather poetic 'need' within the organism, which did not attract much scientific support. For Lamarck the purpose of life was an unfolding of divine will on earth.

His theory is essentially contained in two basic laws. The first states that 'organs, thus species, change in response to a need created by a changing environment'. The second law states that 'such change was passed through the hereditary mechanism to the offspring'. The theory is generally referred to as 'the inheritance of acquired characteristics'.

Jean-Baptiste Lamarck was regarded by many as a great biologist. It was his tragedy that, though he was a founder of the theory of evolution as we know it, his central mechanisms conflicted with Charles Darwin's. Lamarck believed the forces behind evolution were God, environment, then the organisms themselves. Darwin said it was natural selection, organism and environment. In that particular struggle for survival Darwin was unquestionably the fittest.

Lamarck's cloudy and mystical style left his meaning unclear, and made it easy for his opponents to distort his ideas. Those opponents came at him from both sides, the traditional creationists on the one hand, and later the selectionists on the other. The slight change of a few key words was all that was necessary to discredit his ideas: in Lamarck's eulogy after his death in a crucial passage the word 'need' (*le besoin*) became the 'wish' or 'desire', so that instead of an organ changing in response to a need created by changing conditions, it was seen to be moved by some mysterious wish to change. After Lamarck's death, T. H. Huxley accused Lamarck of teaching that long-legged wading birds 'developed their long legs by wanting to keep their feet dry'. (As Darwin's daughter Henrietta pointed out, it should surely have been their bodies.)

Darwin himself expressed various opinions about Lamarck. He praised him for opening people's minds to the concept of evolution; in an early notebook he refers to him as 'a source of inspiration endowed with the prophetic spirit in science, the highest endowment of lofty genius', and in *The Origin of Species* as 'this justly celebrated naturalist'. But in a footnote to that work he condemns his 'views and erroneous grounds of opinion', and in a letter calls them 'veritable rubbish'. From there it was a short step to dismissing him altogether.

The main problem was that Lamarck did not propose any convincing mechanism through which the environment could influence hereditary change.

It was Robert Malthus who put Darwin on the road to his solution. Malthus's *Essay On The Principle of Population as it Affects the Future Improvement of Society* was read by Darwin 'for amusement'; the effect was considerable. The simple point argued in Malthus's *Essay* is that unless it is held back by some restraint, the human (and animal) population will always grow faster than the supply of food. The restraint may be a 'moral' one, as Malthus called it, meaning that people choose to have fewer children, but without that voluntary restraint the population will inevitably increase at an accelerating rate, the food supply will also grow, but more slowly, and in the end hunger and disease will impose their own limit (Malthus, 1798).

Malthus and Darwin recognised that food and space were the ultimate determinants of population size, but neither of them recognised the food and space factor as the initiator, the forward driving force of populations. Neither did they recognise that it was not just the amount of food that matters, but also its chemistry.

However, from Malthus Darwin began to see how any species with even the slightest adaptive advantage over its competing neighbours would be the one that survived. This was the process termed 'natural selection'. Part of the beauty of this two-point scheme was its simplicity: (i) Random mutations occur; (ii) The struggle for existence eliminates all but the best fitted.

On its introduction in 1859 the theory of natural selection offered a clear-cut, easily comprehended and exciting explanation of evolutionary advance which immediately removed the need to speculate further about religious determinism, 'vital forces', 'need', 'wish' or 'will' to progress: it was enough to understand that organisms sometimes made mistakes in propagating their kind, so that chance variations occurred for natural selection to work on. Transcendent forces could be forgotten. Only statistics were needed.

Another factor which made natural selection immediately popular came about with the help of the English philosopher Herbert Spencer. Darwin had only grudgingly agreed with Spencer's suggestion that 'best-fitted to the environment' might be more simply phrased as 'the survival of the fittest'.

It was already the opinion of the Victorians that they were the fittest – the world map was then largely red in colour, the colour of the British Empire. Now their feelings were backed by science: they were the best fitted to rule the world. The struggle for survival, competition, the elevation of the fittest and the subjugation or elimination of the weakest was the way of nature. A sense of perfection caught the imagination of the day, and within five years of publication of the first edition of *The Origin of Species by means of Natural Selection; or the Preservation of Favoured Races in the Struggle for Life*, Darwin had converted many of the leading thinkers of his day.

Many of Darwin's opponents saw *The Origin of Species* as an attack on the doctrine of the church (which it could have been, given Darwin's views on Christianity), and this inevitably made the dispute into a battle between science and religion. It was perhaps the last great battle in a war that had been going on since Galileo's confrontation with the Pope to free science (and the general public) from the church's authority, to allow it to investigate what it chose and to arrive at whatever conclusions the evidence suggested.

Arguments about Darwinian philosophy have continued ever since: but before looking at the critics, let us see how the orthodox view became established.

Adaptation of a living species to its environment through natural selection is now commonly regarded as the force behind evolution. But Darwin himself was convinced there was an additional directive force within the 'conditions of existence' which could operate independently of, or together with, natural selection. Exactly how this adaptation was achieved was at the heart of Darwin's dilemma as he pondered over the problem of the precise mechanism whereby the environment interacted with the organisms themselves. He is rightly known, with Alfred Russell Wallace, as co-originator of the idea of natural selection, but he in fact consistently stated that natural selection worked hand in hand with 'conditions', as many passages in the *Origin* and other writings make clear.

Darwin began by being convinced of the power of conditions. In 1842, as he was first planning his theory, he wrote in his 'sketch' of *The Foundations of the Origin of Species*:

> It must I think be admitted that habits, whether congenital or acquired by practice [sometimes] often become inherited; instincts

influence, equally with structure, the preservation of animals; therefore selection must, with changing conditions, tend to modify the inherited habits of animals.

In 1859, in the first and sixth chapters of the *Origin of Species*, commenting on the origin of domestic species, he says:

> Changed conditions of life are of the highest importance in causing variability, both by acting directly on the organisation, and indirectly by affecting the reproductive system . . .
>
> It is generally acknowledged that all organic beings have been formed on two great laws – 'Unity of Type' and the 'Conditions of Existence' . . . . in fact, *the law of the Conditions of Existence is the higher law* [our italics], as it includes, through the inheritance of former variations and adaptations, that of Unity of Type.

This describes how Darwin felt that the nature of the organism somehow encapsulated within itself all the environments of all its forebears. To him therefore natural selection did not preclude the environment from having a directing role.

He often remarked upon the 'plasticity' of certain plants:

> It is well worth while carefully to study the several treatises published on some of our old cultivated plants . . . and it is really surprising to note the needless points in structure and constitution in which the varieties and subvarieties differ slightly from each other. *The whole organisation seems to have become plastic, and tends to depart in some small degree from that of the parent type*. (Ch. 1)

The idea of plasticity is of particular interest to us in our study, as will become apparent later.

In 1868, also, he gave more thought to external conditions: in a new book called *The Variation of Animals and Plants* (Darwin, 1868), he produced his hypothesis of 'pangenesis'. This theory was that every cell of every tissue and organ of the entire body had minute particles called gemmules (or 'Darwin's genes', as they came to be known). When any cell changed through a change in the conditions, the change would be registered in the gemmules. They would then circulate in the bloodstream and eventually reach the gonads, where the change of the cell would again be registered and so passed on to any offspring. The theory of pangenesis did not last long before becoming extinct. Ironically it perished largely through the research work of Darwin's cousin, Francis Galton, on the blood of rabbits, which revealed nothing resembling gemmules. There being no evidence to back up the hypothesis, within a few years Darwin let it drift quietly away; but the word can still be found in dictionaries.

His hypothesis of 'pangenesis' was a pointer to what he was looking for: not that natural selection was not operating, but that environmental forces could be stimulating change which would later be acted on by selection.

> . . . variability is not a principle co-ordinate with life or reproduction, but results from special causes, generally from changed conditions acting on successive generations . . . apparently due to the sexual system being so easily affected by changed conditions . . .

Although Darwin was acutely aware of environmental influence, he was even more enthusiastic about the total concept of natural selection – with the result that people tended to forget his theories on the environment as a directing force.

After Darwin published, some scientists argued that natural selection was sufficient in itself as the driving and directing force. They of course took the environment, and the food it produced, for granted: these were stable entities which were deemed important, but were not themselves seen to be the directive force in evolution. In this way Darwin's most important conditions of existence ('higher law') came to be seen as an inferior factor when compared to the power of natural selection. A role for nutrition as an environmental and causal factor remained unsuspected.

Later on Darwin felt he had previously underrated the importance of 'conditions', and he expressed dismay that he was often misinterpreted and misquoted on this problem. In the sixth edition of the *Origin*, published in 1872, he wrote:

> I have now recapitulated the facts and considerations which have thoroughly convinced me that species have been modified, during a long course of descent. This has been effected chiefly through the natural selection of numerous successive, slight, favourable variations; aided . . . in an unimportant manner . . . by the direct action of external conditions, and by variations which seem to us in our ignorance to arise spontaneously. It appears that I formerly underrated the frequency and value of these latter forms of variation, as

> *leading to permanent modifications of structure independently of natural selection* [our italics]. But as my conclusions have lately been much misrepresented, and it has been stated that I attribute the modification of species exclusively to natural selection, I may be permitted to remark that in the first edition of this work, and subsequently, I placed in a most conspicuous position – namely, at the close of the Introduction – the following words: '*I am convinced that natural selection has been the main but not the exclusive means of modification*' [our italics]. This has been of no avail. Great is the power of steady misrepresentation; but the history of science shows that fortunately this power does not long endure.

The point could not be made more categorically: 'natural selection has been the main but not the exclusive means of modification.' As to what the other means were, Darwin clearly stated that the conditions represented the 'higher forces'; throughout his life he did not doubt that this other force existed, but failed to come up with a mechanism. Indeed, he kept on revising his great work and seeking some mechanism within the environment, as well as natural selection, that would drive evolution.

It was Professor August Weismann of the University of Freiburg who brought the axe down on Darwin's idea that the environment was a significant power. He extracted natural selection from Darwin's thesis and presented it as a single, pure force. In the late 1880s he published his theory of what he called 'the continuity of the germ plasm'. This theory was a reply to Darwin's theory of pangenesis, which some thought to be veering dangerously close to the heresy of Lamarckism. Plasm is simply the living material of a cell, and the distinction Weismann was making was between the 'germ plasm' of the reproductive cells (in mammals, the male sperm and the female ovum) and the 'somatoplasm' which made up all the cells and tissues of the rest of the body. Speaking of the germ plasm Weismann says, 'This substance can never be formed anew; it can only grow, multiply and be transmitted from one generation to another (Weismann, 1893).'

Weismann believed there was 'an absolute gulf' between the two: it became known as Weismann's 'isolation theory'. The somatoplasm of the parent could not change the germ plasm, under any conditions whatsoever, because of this total insulation. 'It produced the germ plasm, fed it, nurtured it and acted as its vehicle, but it could never modify the instructions that it contained. Encoded in the germ plasm was the complete blueprint for the next generation, which would in turn pass that blueprint on, wholly unaffected by the biochemistry of its somatoplasmic host.' (Bowler, 1983) This was the foundation stone of genetic determinism.

In 1881 Wilhelm Roux proposed a new development with his theory of 'intra-selection', in which he explained how natural selection could be seen to be operating on separate parts of the organism, as well as on the organism as a whole.

In his paper on 'The Effects of External Influences on Development' in 1894, Weismann took the idea a stage further by imagining that, 'natural selection was directed by competition for nourishment among "character units" of the germ plasm, [so] change would move in the direction favoured by selection (Weismann, 1894).'

We shall show how nutrition is not a passive but a causal factor in changing gene expression. Of course, nearly a century ago there was no knowledge of these factors. But it was through this conception (i.e. the isolation theory and intra-selection) that Weismann managed to fix the concept of 'selection', thereby dumping the rest of Darwin's concern about conditions in the rubbish bucket. Weismann's theory was the end of any drift towards Lamarckism. It was the most explicit possible refutation of the inheritance of acquired characteristics. Even Julian Huxley (who, like Weismann, knew a lot about animals but little about plants) in his *Evolution: The Modern Synthesis*, says:

> Although the distinction between soma and germ plasm is not always so sharp as Weismann supposed, [his] conceptions resulted in a great clarification of the position. It is owing to him that we today classify variations into two fundamentally distinct categories – modifications and mutations . . . Modifications are produced by alterations in the environment (including modifications of the internal environment such as are brought about by use and disuse), mutations by alterations in the substance of the hereditary constitution (Huxley, 1942).

Weismann and his followers argued that, as many of these 'modifications' appeared to be non-adaptive (non-adaptive means when an organism changes in a way that does not confer advantage), and as the Lamarckian explanations were meant to apply only to changes which were adaptive, there was therefore no Lamarckian effect.

Weismann further supported his own theory by insisting that if the inheritance of acquired characteristics were a fact, then it should also include the inheritance of mutilations. He must have known that centuries of docking lambs' tails did not produce tail-less sheep, but still proceeded to cut off the tails of 22 generations of mice, over 900 in all, to prove that the trait of tail-lessness would not be passed on to the offspring. (Someone remarked that 'he might just as well have studied the inheritance of a wooden leg'. Koestler, 1971)

It seems that Weismann ran this project to disprove pangenesis, which, had it been true, would have had to include the inheritance of acquired mutilations. In the 1880s there were a number of experiments done to see if regrowth after an amputation caused hereditary change: some of these were quite interesting. Did he cut the tails off mice because he knew he was on firm ground? Whatever his motives, he succeeded in slipping in the inheritance of mutilations as an added requirement for the definition of Lamarckism.

Weismann's theory was definitive in that mutations were thereafter understood to be the stuff that evolution was made of: environmentally induced modifications were made to look insignificant because in laboratory research they proved not to be a form of hard, rigidly fixed heredity which would run true for long periods of time independently of changed conditions. They were therefore not considered capable of directing the long term course of evolution. This view was not entirely shared by the field biologists, who usually saw the environment alter more gradually than the change engineered by the laboratory scientists.

Weismann's theories were vigorously challenged at the time, but successfully defended. The 'all-sufficiency of natural selection' (the title of a later article by Weismann in reply to Herbert Spencer's paper 'The Inadequacy of Natural Selection') was not at all self-evident to many of his contemporaries: nor would it have been to Darwin, who was now dead (Spencer, 1893).

In fact, although Weismann's theory represented what later came to be seen as mainstream neo-Darwinism, many scientists were still dissatisfied. The decades between the 1880s and the mid 1930s were filled with passionate debate. Some thought Darwinism would not survive – this period became known as the Eclipse of Darwinism. Peter Bowler describes the often bitter decades in his book *The Eclipse of Darwinism*:

eal sense of hostility between the two sides, with little
ation. They read different publications and attended
ences. The field biologists and palaeontologists felt
vidence for the formative effects of the environment
arch: but the laboratory scientists had had enough
y endless argument and wanted to get on with
rd facts of heredity and variation (Bowler, 1983).

turning points in the evolutionary debate came
rediscovery and new appreciation of Mendel's
r on genetic inheritance, *Experiments in Plant*
had originally been published in 1865, but in
following the publication of *The Origin of Species* it had gone unnoticed. Now, after thirty-five years of obscurity, it suddenly emerged as the foundation of the modern science of genetics. Although a number of scientists since have thought Mendel's statistics were rather too tidy for comfort, his ideas were nevertheless seen as proof of the correctness of Weismann's theories, and the latter's definitions gave Mendel's system of heredity greater validity.

It would have saved Darwin himself some headaches if he had known of Mendel's investigations. Before Mendel it had been generally believed that the characteristics of the parents blended together in the child. Over a number of generations any particular characteristic was likely to be progressively diluted and eventually to become undetectable. What Mendel showed was that the 'lost' characteristic was still there, carried on a recessive gene. If two parents came together, each of whom had that gene, then the characteristic would again show up in their child. Darwin and others had put this phenomenon down to 'atavism', an unexplained reversion to the character of the grandparent or great-grandparent. Mendel not only explained the mechanism but gave it a mathematical basis. However, he also had some variables in his results: for example, when looking at smooth-coated and wrinkled-coated peas there was a proportion which was clearly neither one nor the other. These he removed, so his figures looked incredibly neat. The percentage of the 'neither/nors' would have confused those figures. Could they be seen as examples of a more malleable type of heredity, working hand in hand with hard heredity?

Mendel's work showed that change of genetic structure (genotype) did not necessarily produce change of the physical form (phenotype). As Bowler points out, this made it easy for people to believe that phenotypic change could not change the genetic structure. Consequently supporters of Mendelian ideas enthusiastically accepted Weismann's theories.

Mendel's was a theory of heredity mechanism, not of evolution, but it was one that any evolutionist would be forced to consider. His investigations had been carried out on peas but his laws held good not only for all other plants but for animals as well, and it was generally thought that any theory of evolution that did not fit with his results could no longer be taken seriously. From this time onwards all evolutionary change had to be because of regular fixed genotypic change. No one looked for a more malleable form of heredity that required no change of structure of the genotype – simply because change of that kind had been labelled 'modification' by Weismann and was therefore unimportant.

One early attempt to incorporate Mendel's laws into an evolution theory was the Mutation Theory put forward by Hugo de Vries in 1906. His idea was that evolution could have proceeded by a series of leaps ('saltations') and not, as Darwin believed, through the slow accumulation of small changes. But, like other theorists before him, he could provide no mechanism for the 'jumps'. Why and how did these saltations occur? Indeed, how could such a massive and co-ordinated change occur suddenly? Chance was an unfortunate, if convenient, choice of words by Darwin. What did it mean? This is something that will be explored in later chapters.

The application of higher mathematics to genetics from around 1917 to the mid 1920s led to another startling discovery. Based on the work of the statistician R. A. Fisher and the biologist J. B. S. Haldane in England, and Sewell Wright in the United States, mathematical biology and genetics were successfully crossed giving birth to the science of mathematical population genetics which showed that when the principle of natural selection was applied to theoretical models, the effects of even small genetic changes at population level would result in the sort of overall change that Darwin himself originally spoke of. Such changes could also be interpreted in Mendelian terms. These discoveries were to unite Mendel's ideas with Darwinism, which began to revive from its 'eclipse'.

The emergent schools of molecular biology and genetics joined forces with the neo-Darwinism of Weismann, recognising the extraordinary power of natural selection, but altering its role. This coming together of previously warring factions became known as the Modern Synthesis. Natural selection, they explained, works at population level, on the 'gene pool' of the species, which is continuously being replenished by small changes attributable to mutation, recombination, migration, isolation, etc. It is on such changes that selection then acts. Struggle, competition and chance continued to drive the process. Stimulation from nutrition was not considered.

The Darwin–Weismann–Mendel line of orthodox thinking became finally cemented through Watson and Crick's discovery of the mechanics and replication of DNA.

Having seen how the orthodox line on evolution theory developed and became conventional wisdom, let us now take a look at the critics and what they objected to. Controversies have continued to surface in many different forms. One of the original disputes was between Darwin and Alfred Russell Wallace over the higher emotional faculties of man. Wallace could not believe that altruistic or idealistic behaviour could simply be the result of natural selection. The recent theory of 'sociobiology' associated with W. D. Hamilton and his followers, E. O. Wilson, Richard Dawkins and others, claims to provide an answer. This school sees all evolutionary progress as a competition for survival not so much between species as between particular genes; the 'selfish gene', in Dawkins's memorable phrase, and genetic determinism lie behind it all – species are merely seen as carriers of particular genes. Going on from there, the sociobiologists claim that natural selection will favour an inherited pattern of altruistic behaviour if it benefits the group as a whole and so makes the survival of the genes it carries more likely. Once again, this has led to a heated and emotional dispute, with some oppponents of the theory going so far as to assert that it could encourage fascism.

There had, of course, been similar objections to the original concept of Darwinism. 'Red and raw in tooth and claw', it appeared to some to be a philosophy which lacked compassion for the weak; one which was loaded toward the mechanistic. It was because of such emotional responses, as well as the feeling that natural selection was not on its own a sufficient explanation of the driving force in evolution, that Darwinism went through its period of despondency and rejection around the turn of the century.

Well into the first quarter of the new century the unorthodox line of dissenting environmental theorists were still objecting, experimenting and making themselves heard. During these decades a number of different theories were developed in attempts at totally new approaches. In 1865 Carl Naegli had forced Darwin himself to concede that the widespread existence of non-adaptive characters represented a major problem for his theory.

The idea of orthogenesis (literally meaning 'straight-line evolution'), whereby organisms were pre-programmed to develop in a particular direction, was proposed by Wilhelm Haake in 1893. Groups of species like the giant reptiles could come into existence together, sharing a common birth, each going through individual stages of development, youth, maturity, old age and senility before final extinction.

An interesting idea was produced in 1940 by Richard Goldschmidt, one of the few front-line geneticists who seriously considered orthogenesis as an alternative to Darwinism. Starting from the idea that only the outcome of those mutations that fitted the existing conditions would be seen, he suggested that this could limit them to a single direction, controlled by the direct environment of the developing fertilised egg ('the surroundings of the primordium in ontogeny'). This, he explained, 'is orthogenesis without Lamarckism, without mysticism, without selection of adult conditions' (Goldschmidt, 1940).

The problem of the orthogenesists was that they could not supply a reason for the unavoidable passage to extinction. What was this force within the organism that compelled it to develop in a certain way? It does not seem to have occurred to them that it could be related to patterns of nutrition or the exhaustion of nutrients from over-worn environments. Without some such mechanism the theory was unconvincing. James M. Baldwin tried to put greater environmental concern back into Darwinism with his theory of 'organic selection' (commonly known as 'the Baldwin Effect') which proposed that species could change shape or form as a direct effect of new behaviour patterns; but, ironically, since he insisted that the change was randomly produced, it became seen as a purely Darwinian theory.

But environmental experiments still continued – sometimes

taking over ten years to conduct. They were designed to demonstrate that characteristics could be produced by environmental change, and be reproduced in the offspring. If this research proved positive, external conditions, as Darwin had originally thought, would be seen as a factor as significant as natural selection.

Many of these environmental experiments did stimulate change which turned up in successive generations. Where they were all judged to have failed was that animals reverted to their previous state when replaced in their ordinary surroundings. The change that was being looked for was, of course, the permanent, rigidly fixed type of heredity, with stability independent of environmental change. After Weismann's definitions, any such malleable form of heredity was merely a 'modification' which did not cut much ice.

Paul Kammerer, a zoologist from Vienna, allegedly produced interesting changes, particularly with salamanders, whose colouring and birth procedures he managed to change through persuading them to breed in unusual environments (something no one since has managed to do). His Ciona experiments were repeats of those of former researchers, yet he was run to ground by indignant academics when he claimed to have proved Lamarckism (Koestler, 1971). His mistake perhaps was to try to fit his own research results into an already confused area, rather than trying to come up with a theory of his own.

Although Kammerer's tragic suicide brought decades of research to a standstill, it did not prevent recurring criticism of natural selection as the sole mover of evolutionary momentum: such criticism has persisted up to the present day.

In 1942 C. H. Waddington took up the argument in an article in *Nature*:

> The battle which raged for so long between the theories of evolution supported by geneticists on the one hand and by naturalists on the other, has in recent years gone strongly in favour of the former . . . [but] . . . if we are deprived of the hypothesis of the inheritance of the effects of use and disuse, we seem to be thrown back on an exclusive reliance on the natural selection of merely chance mutations. It is doubtful, however, whether even the most statistically minded genticists are entirely satisfied that nothing more is involved than the sorting out of random mutations by the natural selection filter (Waddington, 1942).

In 1956 C. P. Martin attacked Weismann's basic theory of isolation between germ plasm and somatoplasm as 'an intellectual artefact which corresponds to nothing that exists in nature'.

Sir Peter Medawar, past Director of the Medical Research Council and winner of the Nobel Prize for Medicine in 1950, was strongly critical of the Darwinian approach, asserting that it uses inductive reasoning leading to a form of 'self-deception'.

It is a common criticism that the conventional theory of evolution works backwards from the present, using a form of inductive logic which imposes individuals' ideas on how things happened based on personal value judgements. The physicist Professor H. Pagels, Executive Director of the New York Academy of Sciences, quotes Einstein on the subject: 'Einstein thought it a mistake to project our human needs into the universe because he felt it is indifferent to those needs' (Pagels, 1983).

Today, Gordon Rattray Taylor summarises the still unexplained problems of orthodox Darwinian theory in his book *The Great Evolution Mystery* (Rattray Taylor, 1983):

> The evidence I have presented is only a sample from a mass of data, but it is enough, I suggest, to prove that natural selection is insufficient to explain all the features of the evolutionary story, and to make it necessary to consider quite seriously the possibility that some directive force or process works in conjunction with it. I do not mean by that a force of a mystical kind, but rather some property of the genetic mechanism the existence of which is at present unsuspected . . .
>
> We have seen at least a dozen areas where the theory of evolution by natural selection seems either inadequate, implausible or definitely wrong. Let me briefly summarise them:
>
> 1. The suddenness [sudden to palaeontologists means happening in a million years] with which major changes in pattern occurred and the virtual absence of any fossil remains from the period in which they were alleged to be evolving.
> 2. The suddenness with which new forms 'radiated' into numerous variants.
> 3. The suddenness of many extinctions and the lack of obvious reasons for such extinction.
> 4. The repeated occurrence of changes calling for numerous co-ordinated innovations, both at the level of organs and of complete organisms.
> 5. The variations in speed at which evolution occurred.
> 6. The fact that subsequently no new phyla have appeared, and

no new classes and orders. This fact, which has been much ignored, is perhaps the most powerful of all arguments against Darwin's generalisation.

7. The occurrence of parallel and convergent evolution, in which similar structures evolve in quite different circumstances.
8. The existence of long-term trends (orthogenesis).
9. The appearance of organs before they are needed (pre-adaptation).
10. The occurrence of 'overshoot' or evolutionary momentum (e.g. how organs, once useful, became overdeveloped, such as the tusks of the sabre-toothed tiger, the antlers of the Irish Elk).
11. The puzzle of how organs, once evolved, come to be lost (degeneration).
12. The failure of some organisms to evolve at all.

The introduction of nutrition and environment in the form of chemistry and physics provides answers to these questions, as we shall see later.

The old problem of reversion which plagued the environmental theorists – the rejection of environmental influence on grounds of reversion – is a puzzle to us. For if an attribute of a species is changed by the environment in a way which supports environmental influence, surely a reversion to the original form, if the environment is changed back to its previous state, is precisely what one would predict on the basis of environmental theory? It would be the condition under which reversion does *not* occur with restoration of the environment which would require an explanation.

The explanation to this puzzle is that no structural genetic change took place: the genetic structure (genotype) remained unaltered and it was only gene 'behaviour' or 'expression' that changed, the genotype regulated, and therefore behaving in a different way.

If the apparently more successful environmental experiments were to be re-assessed, where variation stimulated by environmental change was produced and continued to be produced in subsequent generations, it could collectively be regarded as evidence that change in quality of environmental conditions can and does affect alteration of genetic behaviour. This alteration may look like hereditary change but in fact it isn't. It is a type of *malleable* or *plastic* heredity, caused by environmental stimulation in each generation. Such changes might well be short-lived, but they are short-lived *only* if the environment undergoes changes again in other directions. If the qualities of the environment which stimulated the initial alteration remain constant, or even cumulative, such changes of form or function would stay put or, again, be cumulative.

They could therefore be a more weighty factor in the evolutionary process than is commonly recognised. We refer to this phenomenon as substrate-driven or plastic heredity. At present little importance is attached to such modification as it is not seen as a regular, 'fixed' type of genetic mutation. (Substrate is jargon for the chemicals which are used by, nourish, or influence animal or plant metabolism.)

There are a number of possible ways in which this plastic heredity may work. Firstly, through the suppression of one or more genes; or secondly, through the release of a genetic potential which had previously been suppressed: this could be achieved by environmental – including nutritional – stimulation.

For example: enzyme concentrations in cells may change in response to a signficantly raised, or lowered, intake of a specific nutrient. Enzymes are specialised proteins (proteins themselves are made from amino-acids) which digest and metabolise foods, make ingredients for cell growth, detoxify nasty compounds and deal with spent materials. The amounts of certain enzymes present can be changed by varying amounts of nutrients in the diet, or by hormones, which themselves respond to the nature of nutrients in the diet. In other words nutrients, or chemicals (for that is what they are), can influence the expression of genetic information – i.e. genetic expression can be stimulated or suppressed by larger or smaller amounts of a specific substrate in the food chain.

The dramatic increase in average height in England, of 0.4 inches a decade in the first half of this century, could be a manifestation of such an influence. The time span is, of course, far too short for this to be an evolutionary change of the ordinary kind. It was, rather, a change in the instructions put out by the existing stock of genes (alteration of genetic expression) caused by changes in nutrition.

We know that change can occur in a violent manner from our experience of carcinogenic chemicals in food. We know that mutations can be caused by such mutagens as radioactivity, ultra-violet light, X-rays, certain chemicals and heat. At an early stage in evolution, when the DNA/RNA complex was less well protected, this influence could have been strong.

In more modern living systems, where the DNA is served by

repair mechanisms, it is to some extent protected from such violent perturbations. However, there is little doubt that the introduction of a new chemical into the environment or food chain could, by virtue of its consistent presence, either stimulate change or make possible the existence of a new genetic format.

The important distinction between selection and substrate-driven change is that the mechanism of the latter is active or propelling, whereas 'selection' is passive, effecting a sorting out process after the change has taken place. Such substrate-driven change appears to be capable of directing the course of change both in a progressive forward direction, or in reverse. There is no reason, however, why both forces should not operate together.

The interesting question is this: if there is a sudden change in environmental/nutritional conditions, and if those conditions remain the same over many generations, could this type of malleable change ever become fixed? Suppose, for example, the average height of people in the USA reaches 8 feet by the mid-21st century – will this be a fixed or reversible change? Supposing these people lived together with the mid-19th-century 5-footers, would people consider they were genetically different populations?

Darwin had, of course, pondered on such questions himself. 'Natural selection is a slow process, and the same favourable conditions must long endure, in order that any marked effect should be thus produced (*Origin*: Ch. 7).'

A new form of de Vries's theory is echoed by the theory of 'Punctuated Equilibria' put forward by Stephen Gould and Niles Eldredge (Gould and Eldredge, 1977), in which long periods of stasis are punctuated by 'bursts' of rapid change (rapid, that is, in geological terms). This of course would explain the sudden jumps or discontinuities that occur in the fossil record. (The leap in average height could be an example of such a jump.) Darwin had simply dodged the issue by assuming that the jumps did not happen and sought refuge in the excuse that they were artefacts of an incomplete record. The point put forward by the Punctuationists is that whilst they consider selection an adequate sorting out mechanism, they do not see it as the direct cause or origin of the 'punctuations': clearly there must be another driving force in operation. The Punctuationist view supports a theory of substrate-driven change, for there have undoubtedly been leaps which have been very fast indeed. Later we will see just how fast this can be.

Darwin's own views on this matter are thought provoking:

> If numerous species belonging to the same genera or families, have really started into life at once, the fact would be fatal to the theory of evolution through natural selection. For the development by this means of a group of forms, all of which are descended from some one progenitor, must have been an extremely slow process: and the progenitors must have lived long before their modified descendants (*Origin*, Ch. 10).

This leads us to the other major question on which Darwin has been consistently misunderstood: the role of chance in natural selection. Once again, the message that his more enthusiastic followers thought they heard was not the one Darwin was putting forward. In Chapter 5 of *The Origin of Species* he wrote: 'I have sometimes spoken as if variations were due to chance. This of course is a wholly incorrect expression, but it serves to acknowledge plainly our ignorance of the cause of each particular variation.' It would have been strange if Darwin had believed in random mutations, if by random we mean something which is in principle unpredictable because it is not the outcome of a chain of cause and effect. Such a belief would have gone counter to the whole scientific tradition in which he worked.

And now we come to a very curious fact. In all the controversies over what the causes of diversity might be, no one seems to have paid much attention to the factor in the environment that has the most obvious effect on any organism: food. In the 'Foundations' Darwin did glance at it, considering the possibility that food could manipulate species and citing the case of the Galapagos finches: 'I see no difficulty in parents being forced or induced to vary the food brought, and selection adapting the young ones to it, and thus by degree any amount of diversity might be arrived at (Darwin, 1842).'

However, Darwin did not follow up this hunch, and his followers neglected it altogether. It is this omission that we attempt to make good. The aim of this book is to show how vital a factor food, or the chemistry of food, has been in the process of evolution, from the earliest times when life first appeared until the present day.

In some ways this involves no dispute. Any evolutionist, Darwinian or otherwise, must accept that food is an important part of the environment and that it will have a strong influence in any process of natural selection. A mutant that depends on a nutrient it cannot get is not 'fit' and will not survive; nor will an established species if its essential food supplies run out. Our only comment here is that most evolutionists do not give nearly enough *emphasis* to this obvious truth. They would not deny it, but they seldom see how centrally important it is. The point which is missed is that food is not just food. The milk of different species is not interchangeable; its composition is uniquely tailored to each individual species' varying post-natal requirements. Cows' milk is very much richer in protein than human milk, but human milk is very much richer in essential fatty acids. Food contains a crucial 'qualitative' element which makes possible various specialisations within both the plant and animal kingdoms. In Darwin's day, the fact that there were compositional differences in food, upon which intricate specialisations might depend, was unknown.

Another obvious point is that any new life form must be self-consistent. A mutated lion with the jaws and digestive system of a herbivore is an improbable idea; but if such an oddity ever did occur, it could not survive because important parts of a lion's body are built up from biochemicals that plant foods do not supply. This again is an uncontroversial idea once it is stated, but its wide significance is often overlooked.

Next we can take up Richard Goldschmidt's point that when a new 'genetic blueprint' appears it has first to survive the ontogenic process as the organism begins to grow inside the seed, the egg or womb. 'Thus,' he wrote, 'what is called in a general way the mechanics of development will decide the direction of possible evolutionary changes. In many cases there will be only one direction (Goldschmidt, 1940).' Its own chemistry must match the nutrition that the process supplies. In later chapters we shall see how tight a restraint that can be. For instance, a gene change that programmed a cow to see in the dark like a cat would be nullified by the dearth of necessary nutrients which are needed to build complex eyes before it left the womb.

By introducing the idea of food as a directing force, we can suggest a more comprehensive programme of evolution where the process occurs in five phases within each epoch.

1. The first 'push' comes from the conditions. At any given time a certain set of conditions exists in terms of the richness of chemicals or nutrients available, their concentrations and diversity, the climate, its temperature, humidity, range of variation and so on.

These stimulating conditions make possible the existence of certain types of organism, certain living systems, which then appear in response to the conditions and radiate. Stimulation by 'richness' offers an explanation for Rattray Taylor's first and second problems – it also in part supplies an answer to problem six.

2. Continued stimulation. The conditions are at first likely to be consistent in their stimulation, with nutritional abundance encouraging the new organisms to multiply and their numbers to rise.

3. The 'space/numbers' game. As their numbers rise, space and resources come to be shared more thinly. During this phase Darwinian pressure exerts itself. Those species that develop even slightly advantageous adaptations have the edge on their competitors and are more likely to survive. Once this has happened the system settles down, but as there is little change in the environmental chemistry there is a period of apparent stability. Only co-ordinated change is possible within the framework of the original conditions; answering the fourth of Rattray Taylor's problems.

4. Environmental change. At first the rise in numbers has a barely detectable effect on the environment but eventually, as Malthus described, it accelerates to the point where the environment cannot keep pace. It is unable to replenish its resources as fast as they are used up. The upward surge in numbers and the downward plunge of resources leads to a crisis. The living systems have changed the chemistry of their environment beyond renewal, and by now their struggle for survival has become violent but futile. The fate of the dominant species is sealed. This explains the third of Rattray Taylor's queries.

5. The renewed environmental push. The conditions are now appropriate to living systems of a different sort, and with the disappearance of the old species the way is open for them to develop. (In this context Food equals all impact energies.) This fifth phase is therefore the first phase of the new epoch for the new systems. The cycle has begun again.

The best example of the five phases occurred with the evolution of the animal phyla. The blue-green algae slowly built up the amount of the oxygen they excreted until, toward the end of their Phase 4, it reached a point at which it became increasingly poisonous for them. One-and-a-half billion years and then it was all change.

It took this one-and-a-half billion year period before the pollution of the environment with oxygen created a new richness that made animal life possible. When that happened (Phase 5 of the first epoch, or Phase 1 of the second epoch), the multicellular animal phyla exploded on the scene.

From that time, only 500–600 million years ago, everything else – the succession of the trilobites, graptolites, the fish, amphibia, insects, reptiles, dinosaurs, birds, bees, mammals, man, the whole shooting match – has been crammed into a mere third of the time previously occupied by the blue-green algae. The animal phyla could not have evolved before there was oxygen so the advent of oxygen can be seen to have created this dramatic discontinuity in evolution. This evolutionary jerk, following hundreds of millions of years of consistency, answers the question of whether there were forces other than selection which directed evolution. It also shows how evolution can work at varying speeds.

We could sum up our approach by saying that we see all life forms and their evolution as examples of physics and chemistry in action. The process can be considered as a series of chemical reactions in which genetics, organisms and their environments interact. Variability occurs in both genetics and chemistry but the evidence indicates that, of the two, chemistry is the more coercive. Genetics can only operate within rigidly defined chemical conditions. In this way chemistry can be responsible for discontinuities or jerks in evolution, can stimulate the emergence of a new species or be responsible for the demise of the old. To understand this debate we must begin by questioning what is meant by chance or randomness and by looking at some of the basic chemistry which makes up our bodies. To find the origin of these chemicals we need to go back to the origin of our solar system, and even back to the origin of the universe itself.

2 Birth of a living planet

The conventional theory of evolution works backwards from the present using a form of inductive logic which imposes human ideas on how things happened. The scientific approach attempts to build knowledge progressively, testing the validity of each step. In this way, an ability to predict is developed. We propose to see what happens if we start the evolutionary argument at the beginning, which for us means the beginning of chemistry and life.

The value of starting at the origin can be illustrated by the example of the attempts to reproduce a Stradivarius violin, an instrument which today would change hands for over $1,000,000 whereas a modern violin might sell for $500. The magic of the Stradivarius has withstood the test of time. Several attempts to produce anything which even approaches the beauty and quality of its tone have failed. Recently, a team of Japanese scientists employed sophisticated computers to analyse every nuance of the sound in an effort to find clues from its behaviour as to how to build an equivalent instrument. But their analyses have brought little reward. In Texas, Dr Joseph Nagyvary attacked the problem from the opposite end; he did not study what the violin did but asked how it was made.

Dr Nagyvary obtained the co-operation of Rene Morel of the Library of Congress, who made available five Stradivarius violins so that he could study their structure and chemical composition. He could in no way damage the instruments. Despite this limitation he found three key elements. First, there were many small cracks and holes in the wood which were mostly filled with debris. These acted as a high frequency filter and contributed to the mellowness of the tone. Secondly, he found that many disparate elements were used in the preservation of the wood. Thirdly, Stradivarius had used a natural varnish which, unlike the modern polyester varnishes, made a substantial contribution to the tone.

Through an analysis of how the violins were built, Dr Nagyvary has come closer to reproducing the pure sound of the originals than any one else in 200 years. This was because he attempted to understand how it was built, whereas the Japanese attempted to understand its behaviour.

To a large extent that is what this book is about. In a way, biological thinking on evolution has been like the Japanese approach to the Stradivarius. Observers have looked at plants and animals and said, 'This is what it does and needs, therefore this is how it evolved.'

In a sense we want to look at the wood and the glue of a buffalo and a lion in order to learn why the buffalo has no night vision, and the lion has. We will try to build from the beginning, an approach which will work towards our own time and then beyond to offer predictions of the future. We cannot look at all aspects in biology from a reconstructive position because so little hard data is available. But we can examine certain species and lines of evolution on which we do have information, and so attempt to focus on areas where some firm scientific data does exist.

We have discussed some apparent flaws in conventional evolution theory: our purpose here is to show how chemistry and food can be brought into the discussion to explain certain evolutionary tracks. We wish to show how food and the chemistry of food can provide explanations for the discontinuities and grooves in evolution. This attempt at a reconstruction is a tall order for there is nowhere nearly enough evidence and we would not suggest that our solutions will be the only plausible ones. However, we believe it is necesary to open up another aspect of the evolutionary debate, not only because it says something about mankind's present situation, but more importantly, because it offers a predictive capability for his future.

The beginning starts with chemicals. The chemicals that make up the body were born in a star. Every particle of matter in the delicate tissues which build our bodies, is made of elements that were transmuted into their present form in an unimaginable heat and pressure. We are what we are because of the nature of those elements: from the moment they were first created they contained in themselves all the information necessary to produce life and later to produce that particular form of life that we know as ourselves. The origin of life is inseparable from the origin of chemicals.

The most generally accepted theory on the origin of matter suggests that the universe began 15 and 20 billions years ago with a 'big bang' (Weinberg, 1983). If this theory is correct then the matter of the entire universe was concentrated in a single point. Nothing else existed: perhaps even space and time as we know them did not exist. Then the explosion happened, and all the material that now extends through uncounted billions of miles was flung out into space, propelled by a concentration of the universe's energy.

It is not important for us to ask who lit the fuse and nor do we wish to be involved in the arguments about big bang or continuous creation. What is important, however, is to understand that the stars are basically nuclear furnaces in which the elements that make up our homes and bodies were made.

Throughout space, whirlpools of cosmic dust, mostly hydrogen, gathered into clouds in an expression of mutual gravity. The growing gravitational force drew in more and more until a size was eventually reached at which the mass at the centre was so compressed that nuclear fusion occurred between the atoms. The fusion developed a chain reaction which turned the cloud into a colossal nuclear power plant: a star was born.

To understand this process we need to grasp two things. The first is that when any gas is compressed it heats up, just as the air in a bicycle pump heats up when you squeeze it into the tyre, and some of the mechanical energy of your muscles is converted into heat energy. The second point is that when hydrogen is subjected to extreme temperatures and pressures, its atoms fuse to form helium with release of a large amount of spare energy. This principle is used to make hydrogen bombs.

It is remarkable to think that the warmth we feel on our face from the sun is the warmth from a nuclear furnace. The sun radiates such an immense amount of heat into space that even the tiny proportion that lands on our own planet provides us with 2,000,000,000,000,000,000,000,000 joules of energy each year. To imagine what this means, consider that the small part of this total that is used by the plants is enough for them to fix 300,000,000,000 tonnes of carbon dioxide into their materials.

The vast heat of the sun is difficult to imagine. This immense focus of energy is the manufacturing plant of the universe: energy is condensed into matter. Above the surface of the sun the temperature of the swirling gases is between 6,000 and 10,000°C;

near its centre this rises to 13,000,000°C, and at such a temperature and pressure the alchemist's dream comes true: hydrogen, which is the simplest and smallest element, fuses to make higher elements. The elements themselves are then transmuted and base materials become gold. Indeed, all the elements that exist are built in this way.

If we picture an atom as a central nucleus round which the electrons revolve like planets, then hydrogen consists of a single electron circling a nucleus composed of a single proton. The next step up is helium, which needs two protons and two electrons. So the elements are simply basic energy/matter units with the nucleus carrying a positive charge surrounded by a corresponding shield of negative electrons. If an electron is removed, the remaining part becomes positively charged. The attraction of the electron to return to this positive charge is so strong and the speed of its execution so fast, that a single electron would in one second, travel seven times round the equator to find its positive mate!

The business of condensing energy into matter and fusing more and more matter into atoms built up the array of 105 elements which we now have on our planet. What determines which elements are made is simply the temperature and pressure at which the 'alchemy' takes place: i.e. the conditions. Hydrogen, requiring only one unit of matter, becomes the most easily produced and is therefore fairly common throughout the universe. Carbon needs 12 times as many units, oxygen 16, iron 56, silver 108 and gold 197. With the higher atomic weights, some of the nuclear particles have weight but no electrical charge: they are padding. Others act as glue to hold the bigger atoms together. Indeed, ever since Rutherford 'split the atom' in Cambridge in 1932, atomic physicists have uncovered many types of subatomic bits and pieces.

There is obviously a limit to this process of building bigger and bigger atoms. Once made and cooled, the largest of the atoms, like uranium and plutonium, are unstable and begin to break down spontaneously. The particles which fly out of them are what we call radioactivity. Smaller atoms can also be made radioactive if their nucleus is forced to contain more subatomic particles than is stable. This is done for medical research. To follow what happens to a food or a drug all you have to do is to bombard the molecule at high velocity with subatomic particles. This usually means bombarding the carbon, which increases the weight of its core from 12 to 14 units. At that weight its atom is unstable – bits fly off emitting radioactive rays, so the radioactive 'label' in the drug can be followed as it progresses through or is changed in the body. As these radiations emerge, the atom changes and eventually settles as the smaller but more stable atom. Uranium is one of the largest with an atomic weight of 238. On our planet it is now slowly disintegrating into harmless lead at a weight of 207. In 4.5 billion years (aproximately the age of our planet) our original uranium has lost only half of its activity. It takes time to reverse the alchemy of the stars.

At the beginning of the life of a star it is thought that it is only the hydrogen within it which undergoes thermonuclear fusion. As this process progresses, a time will come when the star begins to run out of hydrogen fuel, which means that it is beginning to use higher and higher elements as fuel. Imagine a situation where the space occupied by 16 atoms of hydrogen is now condensed into only one atom of oxygen, 23 atoms of hydrogen into one of sodium and even 238 of hydrogen into one atom of uranium. Obviously the space originally occupied by the hydrogen has shrunk dramatically and so the internal pressure builds up. As larger and larger atoms are made from smaller atoms the pressure rises more and more steeply.

Suddenly, the star can no longer hold itself together and the whole thing explodes in the most spectacular of all pieces of galactic fireworks: a supernova. Chinese astronomers had the good fortune to witness one in 1055 AD when the Crab Nebula suddenly flared in the night sky. In a sense, a star just before it explodes is like an egg before it hatches. During the incubation, hydrogen has been turning into metals and other elements; towards the end of the gestation there is a rapid growth-spurt which culminates in the birth of new matter to be distributed across space to form new planets or even solar systems.

The conditions – i.e. the amount of hydrogen at the start, the size of the star, its temperature and pressures – involved in this event will determine the mixture of elements that are formed, the energy of the explosion and the distribution of the products through space. Our planet is a spin-off of this type of process: a collection of the waste products of stars or the debris from an exploding supernova.

Although there are today a number of conflicting theories, it was in some such way that our planet came into being, seeded with a rich array of chemicals that ultimately became life. At the beginning, it was a white-hot, fiery furnace itself. The remains of that heat persist today and the interior of the earth is still molten, covered by a solid crust so thin that in proportion to the planet's size it is like a sheet of paper wrapped round a football. Even today, the molten interior may break through the crust, spill molten lava on the earth and throw tons of dust and gas into the atmosphere. With the exception of the few remaining volcanoes and hot springs, most of the earth's skin is today within the temperature and pressure range of carbon chemistry or what is now referred to as 'biological chemistry'.

The white-hot stage was cooler than the centre of a star, and at these more modest temperatures the most important chemical event after the creation of the elements must have occurred – the formation of compounds. Paradoxically, although heat is needed for many chemical combinations, the intense heat inside a star is so great that such combinations are impossible. The heat breaks the elements apart as fast as they try to combine. Under these conditions, the chemistry is the chemistry of atomic particles, not the chemistry of elements. If atoms smash into each other, the velocities are such that all they may do is to fuse to build a bigger atom. It is only in cooler places that elements keep their distance sufficiently to form compounds.

Once again, we see the importance of conditions. As the elements cooled they reached a point at which they could combine with one another. Oxygen and hydrogen can combine with many other elements, and did so, but at different temperatures. At very high temperatures oxygen would combine with silicon, sodium and other metals to form compounds which would have been among the first to condense out of a gaseous phase. The surface of the planet on which you stand is little more than the result of high-temperature reactions between sodium, silicon, oxygen, iron, aluminium, carbon and other common elements.

The commonest element in rocks is promiscuous oxygen. The limestone of the white cliffs of Dover is little more than a combination of calcium and carbon dioxide: calcium carbonate. A cairngorm, the prized, semi-precious stone used to decorate the daggers of the Scottish Highlanders, is simply silicon oxide. An amethyst is the same silicon oxide crystallised inside rock formations and with added elements like manganese to give it the blue colour. Flint is a duller but more compact version of the same. Imagine burning silicon in the presence of sodium and oxygen: the result would be the sands of Miami beach.

Many reactions other than the formation of rocks also took place at that time. Nitrogen, sulphur and phosphorus combined with oxygen to form the nitrates, sulphates and phosphates; at these temperatures, simple phosphates combined together to make pyrophosphates which trapped the energy and pre-dated their later use by living systems as energy stores (e.g. as adenosine triphosphate). Sodium and chlorine fused to form salt. Hydrogen reacted with oxygen to form water, and with nitrogen to form ammonia. It also reacted with one atom of carbon to form methane, with two carbons to form ethane, three to form propane and so on to produce octane and possibly even higher hydrocarbons or oils. It is not an exaggeration to say that anything capable of reacting with anything did so to produce anything that was possible. The only requirement for us to know about it today was that the conditions were suitable for the product to remain.

Few elements escaped compound formation as our fireball cooled. Any surplus hydrogen would have been spun into space along with other light gases like helium. At the other end of the scale, unreacted heavy elements would have crystallised. Iron, with an atomic weight of 56, was in great abundance. It sank through the molten exterior and formed a central core. However, despite its lack of reactivity at today's temperatures, no deposits of pure silicon have been found: it is present as oxides in rocks and sand. So one must assume it was highly reactive at the start-up temperatures. Gold, unreactive under the conditions at that time, remained aloof and pure, isolated from all this chemical excitement by its Garbo-like desire to be alone. The bulk of carbon became oxides or carbides destined to be converted into the skeleton of life.

Free carbon is found only as graphite, compressed by the cooling of molten rocks, or, at higher pressures, as diamonds, which are of such rarity that any one piece over 100g in weight would be priceless. It was the conditions at this early stage which created the wealth of De Beers! In 1988 a red diamond was sold by Christie's. It had 97 points, which means that it weighed just under one fifth of a gram only. Red diamonds are much sought after and so the conditions which created this tiny stone gave it a price tag of $880,000!

This union of elements was not a matter of chance. If we place a number of chemicals in a test-tube and raise them to a certain temperature, we can predict quite accurately not only which compounds will form but also in what proportions. For example, if we put hydrogen, oxygen and lead together the probability of oxygen combining with hydrogen is many times greater than with lead: but if there is enough oxygen, and if we raise the temperature high enough then there is a high probability of lead oxides being formed as well as water.

A modern chemist, to take another example, may want to make nitrotoluene. To do this he will adjust the amount of nitric acid and toluene so that he gets as high a yield of the nitrotoluene as he wants. He is, of course, a well-practised expert. At one time he was a student at university and his first attempt probably produced not just nitrotoluene but dinitrotoluene and even the highly explosive trinitrotoluene (TNT), out of which Nobel made all his money to create the Prize. Given the appropriate conditions, chemicals will combine as much as they are able. It is the conditions which determine what is produced, just as it was the conditions in the original star and in the swirling cosmic dust which gave this planet its original gift of chemicals.

If we had been present at the time the elements were first combining, and if we had been gifted with a perfect knowledge of their relative commonness and of the conditions prevailing, then in the same way that the chemist can predict the outcome of his combination of toluene and nitric acid, so we would have been able to say what combinations would occur in the making of our planet, and how much of each compound there would be. The evolution of compounds, like the evolution of life, follows its own rules and what to individuals looks like chance, is actually an inescapable law.

After its creation about 4600 million years ago, it took 700 million years for the temperature anywhere on our primitive planet to drop below the boiling point of water. This would not have been a steady process. Different parts of the earth's surface would have been at different temperatures, and the temperatures would change as day turned into night, and as season followed season.

Since the atmosphere cooled faster than the surface, the first drop of water would have fallen tentatively to the ground, almost certainly at night, at the poles, in winter. It was then promptly boiled back into the atmosphere. This would have gone on for millions of years, but eventually the heat radiated from the earth's crust could no longer keep the atmosphere hot enough to hold the water as vapour. The rain would then have started in earnest. It would have fallen in quantity at night, with the first onslaught being fought back by the hot surface, until the continued fall of water had cooled the outer crust enough for some of it to settle. Still, the heat below the surface would have been too close and too strong, and when daylight came and added the sun's power the water would have been boiled away once more. Each time water fell it sapped some of the heat of the planet's crust. The mass effect of the downpours would have shaped and eroded the crust starting the process of sedimentation. Meteorites would have slammed into the planet, which was at the time less protected by a high atmosphere. They left craters and each crater formed an ideal shape for a pond of water. At last the time would have come when, first in a few places and later in many, sheets and pools of liquid water finally lay continuously on the surface of the earth.

These first rains would have had another very important effect. The earth's crust was covered with many different chemicals created in the hot phase. Among them would have been large quantities of soluble salts of sodium, potassium and ammonium: the phosphates, carbonates, chlorides, nitrates and sulphates and so on. Each time the early rains fell they were washed down into gulleys, depressions and craters, where they formed hot, supersaturated, muddy salt lakes. Ultimately such rich soups at such warm temperatures created the fertile environment for chemical reactions. The clay in the soups could have acted as a matrix and catalyst for many of the chemical reactions that were taking place, encouraging greater degrees of molecular complexity.

Electricity also contributed to the energetic forces at work. When the temperature fell to the range that allowed rain to form, the thick pall of water droplets, chemical gases, vapour and steam were thrown around the planet by high winds. These winds were caused by the rotation of the planet and the violent temperature range of the surface. The friction between the dense clouds and between the clouds and the planet's surface then generated intense electrical energy, in the same way that pulling off a pullover produces electrical sparks simply because electrons have been pulled out of atoms by the friction. A Wimshurst machine generates high voltage electricity using the same principle of friction with a wheel.

Even the atmosphere at that time would have contained simple carbon compounds which would have equilibrated and dissolved in the clouds and rain. The almost continuous sheet and fork lightning cut through the atmosphere, turning night into day and simultaneously energising the simple carbon and nitrogen molecules dissolved in the rain, making them into complex organic molecules. Even today, Russian work has shown that the reason plants grow better following an electrical storm compared to watering from a tap is that lightning fixes nitrogen as nitric oxide, which dissolves in the rain and provides fertiliser both as nitrate and by dissolving trace elements from soils and rocks. Just under four billion years ago, our planet with its swirling clouds formed a giant Wimshurst machine, discharging electrical storms of great intensity: a machine that collaborated with the heat and the sun's ultraviolet rays to produce the basic chemicals needed for life.

A new phase had begun: the chemistry that had built the chemical compounds and rocks at white-hot temperatures was now building organic chemicals based on carbon at much lower temperatures: these were the predecessors of the first biochemicals. Whilst the origin of the universe had been the heyday of particle physics, the origin of our planet was the phase of inorganic chemistry, which was to be followed by organic chemistry and then biochemistry.

Eventually all the compounds necessary to provide for life had been built. The rains continued to fall, the muddy puddles grew in size and temperatures and concentration subsided. At some point in this transition, the perfect, nurturing levels were reached. The materials were there. The conditions were right. The earth was ready for life to be born.

3 The chemicals of life

If the conditions of the big bang created matter, and the conditions in stars and supernovas built the elements, was the origin of life similarly dependent on conditions? Our answer is yes, but to explain why, we need to share with the reader a knowledge of the chemicals which are used to build our tissues so that we are speaking the same language. To discuss the origin of life, it is as necessary to have this knowledge as it is for an architect to know about the materials for the building he is to design and the conditions it has to tolerate.

Firstly, our form of life is based on chemical reactions of carbon compounds in water. These reactions will only take place in a very narrow temperature window between 0° and 100°C. Since the temperature range of the universe is from absolute zero at −273°C to above 1,000,000°C we can see that our own conditions occupy a very small window at the bottom end of the range. These conditions are what gave us life and made our evolution possible.

We know that plants and animals are made of chemicals. Is there a difference between these 'biochemicals' and ordinary chemicals? For many centuries life seemed so mysterious, it was generally believed that some other influence – some 'vital force' – must be at work. The vitalists believed that an absolute discontinuity existed between the material of living creatures and all other matter. It was held that ordinary chemicals could never be made to produce biochemicals – only a living cell could do that. This belief was destroyed in the 19th century by chemists who broke down organic molecules to their component chemicals and others who made organic chemicals from simple laboratory chemicals. The gulf had been crossed. The gap between the living world and its environment was closed.

Since then large numbers of biochemicals have been synthesised and Hoffman la Roche makes vitamin C by the ton. Indeed, there is no reason to suppose that we could not manufacture all the complex biochemicals that appear in living forms: indeed, if we were prepared to invest enough time and money we could map the whole sequence of the human genetic molecule and then make it. There is nothing magic in that. Once again it is a matter of conditions.

To understand the role of chemistry in life we should look at the nature of the interactions between chemicals. Are chemical compounds based on chance or are they determined? To answer this we must look again at the atom as a nucleus with electrons circling it. (A modern subatomic physicist exploring the subtleties of quantum theory would find this model oversimplified, but it is close enough for our purposes.) The nucleus of the atom is formed of protons. These are more massive than the electrons and carry a positive electrical charge whereas the electrons carry a negative one. As with the poles of a magnet, opposite charges attract one another, similar charges repel. Two electrons or two protons will therefore try to move apart; a proton and an electron will pull each other together.

Atoms are built in a star by fusing more and more protons, together with other more elaborate particles, into the nucleus. As more and more protons are packed in, so more electrons are drawn in to circulate around the outside. These electrons are packed in shells at increasing distances from the nucleus. Depending on how effective these shells are at screening the positive nature of the nucleus, the atom of one element may have a tendency to be more positive whilst another may be more negative. When atoms of an opposite nature meet, they are attracted to each other and join up to form a single unit. They will only actually join together if one partner is willing to give an electron and the other to receive in a mutual manner whereby both end up sharing the electron. Elements can share more than one electron and so be joined by more than one bond.

This means that there are very precise rules governing what will join up with what, how many atoms of one element can join with another, and the strength of the union. Some elements, like oxygen and hydrogen, are strongly attracted to each other whilst others require persuasion. This means that the conditions will be a determinant of the types of unions that will take place.

Carbon, the basis of our life, is unique. Most elements make only one or two bonds and this strictly limits the complexity of the compounds they can form: carbon can make four. Furthermore, each of the four can be either a giving or receiving bond, opening up a wide range of opportunities. Because carbon can both give and receive it can join with other atoms of carbon, and at the temperatures of our planet, carbon–carbon links can form daisy-chains, rings and multiple rings. It can react with hydrogen, oxygen, phosphorus, sulphur and many other elements: indeed it can join with four different elements at the same time and they can join with other elements or more carbons leading to the incredible complexity that became life.

Methane is one of the simplest compounds to be formed from the union of carbon and hydrogen: it is made from one atom of carbon and four of hydrogen. If the conditions could create this simple unit, it is certain they would produce carbon–hydrogen combinations with two, three or more carbons strung together, probably resulting in some oily mixtures to accompany the methane gas. Fig. 1. illustrates how the basic chemicals of life are simply different arrangements of carbon, hydrogen, oxygen and nitrogen.

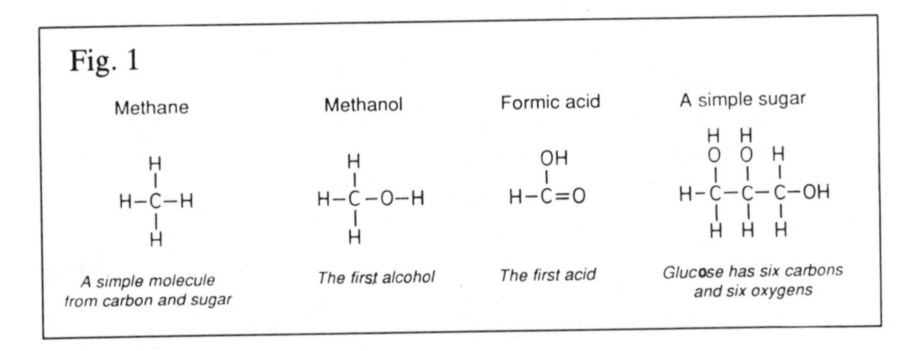

Oils are simply built up from methane by daisy-chaining more carbons (*see* Fig. 2).

Fig. 2

● = Carbon   O = Hydrogen

Heavy oils are simply long chains of carbon and hydrogen – the longer the chain the heavier the oil. We use the short chain lengths as fuel, the long chain lengths as engine oil and the even longer ones as gear-box lubricants (*see* Fig. 3). All compounds which consist only of carbon and hydrogen are known as hydrocarbons.

Fig. 3

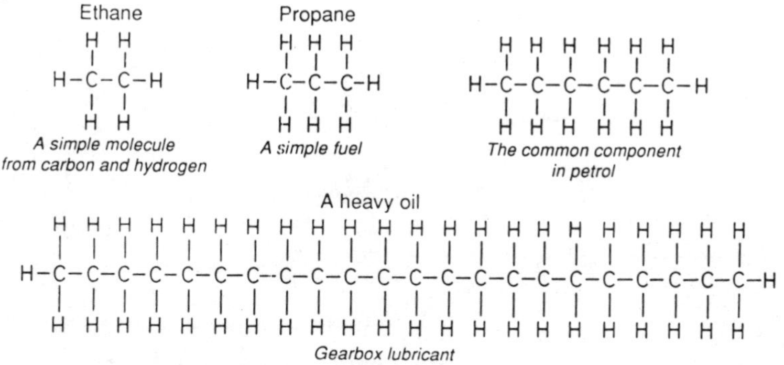

```
   Ethane              Propane                 
    H H                 H H H              H H H H H H
    | |                 | | |              | | | | | |
  H-C-C-H             H-C-C-C-H          H-C-C-C-C-C-C-H
    | |                 | | |              | | | | | |
    H H                 H H H              H H H H H H
```

A simple molecule from carbon and hydrogen — *A simple fuel* — *The common component in petrol*

A heavy oil

```
  H H H H H H H H H H H H H H H H H H H H H H
  | | | | | | | | | | | | | | | | | | | | | |
H-C-C-C-C-C-C-C-C-C-C-C-C-C-C-C-C-C-C-C-C-C-C-H
  | | | | | | | | | | | | | | | | | | | | | |
  H H H H H H H H H H H H H H H H H H H H H H
```

Gearbox lubricant

A simple alcohol is formed by using the same atoms plus one of oxygen. The higher alcohols simply contain more carbons in the chain (*see* Fig. 4). Methanol is dangerous to drink: it destroys the optic nerve. But to change methylated spirit (which is what methanol is) into the kind of alcohol that goes into wine or gin, we need only add another carbon with its hydrogens.

Fig. 4

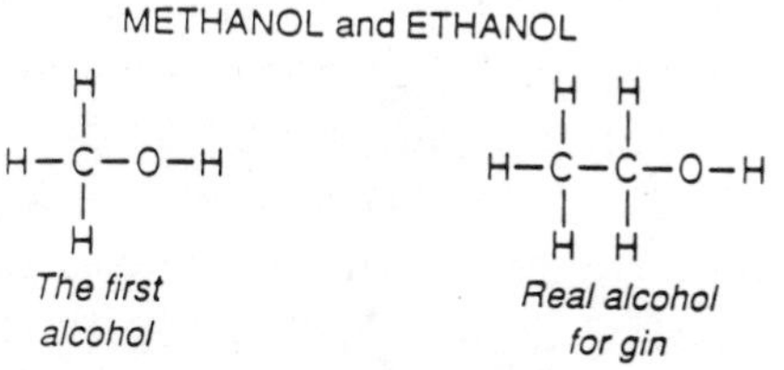

METHANOL and ETHANOL

```
    H                 H H
    |                 | |
  H-C-O-H           H-C-C-O-H
    |                 | |
    H                 H H
```

The first alcohol — *Real alcohol for gin*

This use of just one extra carbon atom again transforms the nature of the compound. With one oxygen you have the basis of Scotch whisky, add another and you have an acid (vinegar), and add a nitrogen and you have an amino-acid – the basic building brick of protein. Add some more carbons to the vinegar and you have fatty acids. Fig. 5 shows the composition of vinegar, the simplest amino-acid (glycerine) and a more complex amino-acid (alanine).

Fig. 5

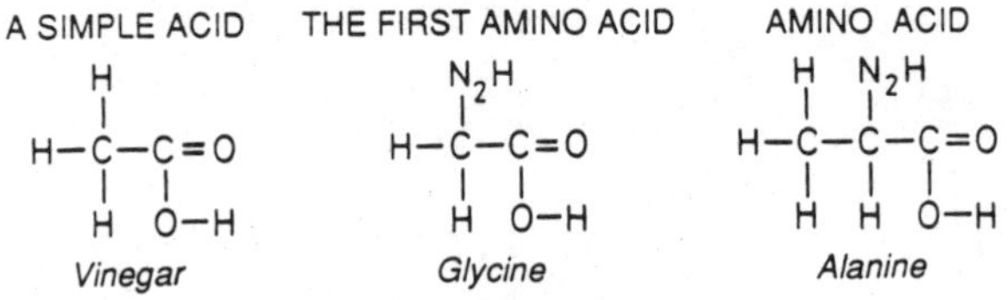

```
A SIMPLE ACID      THE FIRST AMINO ACID      AMINO ACID
     H                  N2H                   H  N2H
     |                  |                     |  |
 H—C—C=O            H—C—C=O               H—C—C—C=O
   |  |               |  |                  |  |  |
   H  O—H             H  O—H                H  H  O—H
  Vinegar             Glycine               Alanine
```

So here we are approaching the boundaries of life, for it is the combination of different amino-acids that form proteins, the building blocks of living tissue, and the sugars that form carbohydrates and the fatty acids that make fats and, ultimately, biological membranes.

All carbon-based molecules are known as organic compounds, and from this very brief expedition into organic chemistry, the sheer number of different compounds that can be built up using carbon and a few other elements can be appreciated. The chemicals of life are simply different arrangements of carbon, hydrogen, oxygen, phosphorus, nitrogen and sulphur. We can also see what a difference can be produced in the nature of a compound by an apparently small change in the arrangement of the atoms that go into it. Methanol is poisonous but many drink and enjoy ethanol. It is no wonder that living systems are easily disrupted or that they will only arise in finely adjusted conditions.

In providing these conditions there is one vital material which is not an organic chemical at all. Like carbon, it is central to life on this planet, it makes up about 80 per cent of the substance of living things and without it they could not exist at all. It is water.

Once again, the unique properties of water are a function of its structure. Its oxygen atom is much bigger than the hydrogens, so the shared electrons spend more time around the oxygen than the hydrogens. This means that one end of the water molecule is positive and the other negative. Any other parts of the molecules of other compounds that carry either negative or positive charges will be attracted towards one or other end of the water molecule.

Hence any charged substance will dissolve in water, simply as a result of meshing in with the + and − charges: its molecules happily fit into the water, each one attaching itself to the appropriate positive or negative end of a water molecule whose positive and negative ends are seeking to be matched.

When you shake up oil and water they at first seem to mix, but soon little bubbles of oil appear as the water molecules are drawn together by the mutual attraction of their + and − charges. The oil is totally neutral, has no charge to play with and simply does not fit into the game. It is excluded and floats to the top. This is why it is no use trying to put out an oil or petrol fire with water.

How then can we wash oil off our hands with water? We can't. What you need to do is to find a substance which is half neutral and half charged: one part of the molecule will possess no electrical properties and so fit into the oil and the other end will be electrically potent and fit into the water. This is precisely what soaps are. They have an acid group at one end of a hydrocarbon. The acid group is strongly charged and shoves its head into the water leaving the neutral tail out of the water but sunk into the oil! Detergents are similarly constructed, as are the walls of a living cell.

Just as the high temperatures of the fireball oxidised silicon to form sand and rock, so the later, cooler temperatures and conditions contrived to produce the organic chemicals from the simplest, described above, to those that were far more complex. It could not have been otherwise. N. W. Pirie commented that it would have been more surprising if it had proved impossible to make such 'biochemicals'; indeed it is very likely that wherever in the universe there are appropriate conditions and raw materials, these types of biochemicals will be formed.

The possibility that the basic 'life chemicals' can arise spontaneously without life was tested experimentally by A. I. Oparin in Russia in 1949 and by S. L. Miller in 1953 in the UK. A powerful electrical spark was repeatedly discharged through a sterile atmosphere of methane, ammonia, hydrogen and water vapour, and after running the apparatus for a week they were able to identify four simple amino-acids in the water. Many other investigators have confirmed the results of this experiment and in addition have been able to observe a number of other organic compounds, including two known as purines and pyrimidines which are the building blocks of DNA.

'Biochemicals' have actually been detected in meteorites arriving from outer space but that does not mean that they have come from a source where there is life. All this evidence demonstrates is that the 'biochemicals' which we associate with life can arise spontaneously and it is actually quite a simple business to generate them.

Professor S. F. Fox took the investigation further. He was able to show how mixtures of amino-acids, could, simply by heat, be induced to link together in long-chain molecules: a process known as copolymerisation which is a process of daisy-chaining smaller molecules together in a line. The significance of Fox's observation is that it is by just such a process that amino-acids joined to form proteins and eventually to form life.

An interesting sidelight is that silk is basically a daisy-chain of simple, naturally occurring amino-acids strung together by the silkworm. Silk is in a sense a protein. In the 1930s Wallace Carothers discovered that you could daisy-chain molecules which were similar to amino-acids but had a different arrangement of the atoms and produce a polymer with somewhat similar properties: he had invented nylon. No process of random mutation has produced an animal with nylon components.

These examples provide an insight into the question of chemical determinacy versus chance. If chemistry is based on chance then life is based on chance. If chemistry is deterministic then life is the same.

In practice, chance plays its part. There is a chance that two molecules will collide. But the result, whether a new chemical product or a non-event, is determined by whether or not these electrical structures provide attractive or repelling prospects. When there is a choice it is always the thermodynamically most stable form which will be made in the greatest amounts; this principle is all you need to convert chance into prediction.

The remarkable similarity of certain genetic codes for crucial aspects of biology (e.g. cytochrome C has been sequenced in over 42 different systems and exhibits only minor differences), presents the same evidence for thermodynamic determinacy at the more complex level of protein and nucleic acid sequences. In other words, certain sequences may appear more frequently for simple reasons of chemistry rather than chance.

In a similar way, drug companies which used to base their search for active compounds on trying everything in the book, a process of trial and error, are now turning to chemical logic. Drugs for use in medicine were originally developed from plant products with known physiological action (curare, used for poison darts by South American Indians, is now used during anaesthesia). When native practice and folklore was exhausted, the drug companies had to search the natural world and laboratory shelves for useful compounds. Today active centres of target enzymes are being mapped by spelling out the sequence of the protein. Computer graphics are then used to portray the active centre in its flexible, three-dimensional shape together with the positioning in space of all its electron-active components. The human eye and brain, together with the computer, design a chemical of the right shape and electrical charges, in the correct spatial alignment, to clamp into the active centre and block the action. There is nothing random about this approach which is pure science-building knowledge. The three-dimensional approach is currently being used to attack the AIDS virus with a strong probability of success.

Up to this point, we can see the life-forming process as a logical sequence. The conditions arose in which chemicals reacted with each other to produce organic molecules of increasing complexity and eventually biochemicals. The possibility that chance goes much further than a single encounter is not supported by the evidence. There were not a number of ways in which the chemicals could unite. The options were limited. If there had been various paths open to chemical evolution then surely in the billions of years since then some of the other possible chemical systems would have appeared and been used in the random collection of new life forms that neo-Darwinian theory predicts. Yet they have not done so.

It is one of the most striking facts in the whole of evolution that the basic chemistry of all life-forms has not changed. Guinea-pigs need the same vitamin C, vitamins B1 to B12, folic acid, vitamin A and D as we do. The same amino-acids are used in their proteins. The same chlorophyll is used in the green leaves of all plants and the same haemoglobin in the blood of different animals.

This persistence in using certain chemicals and rejecting others is hard to explain in neo-Darwinian terms, and it is particularly striking in the way organisms treat what are called isomers, or 'mirror-image' molecules. These are molecules of identical composition but arranged differently in space.

If you put your right hand on a table, palm up, you can put your left hand on to it and the thumbs and fingers of the two hands will all match. The same thing happens if you put your hand flat on to a mirror. Hand and image are the same shapes, but as mirror images. However much you turn a left hand round it will never become a right hand – just as the mirror image has a different spatial arrangement to the original.

In the same way, the three-dimensional shape of carbon molecules can be made in mirror-image forms – the basic building materials of life can be either left- or right-handed.

The two mirror-image forms of a molecule – its isomers – are in chemical respects identical. They will take part in the same reactions to produce the same results, and if we make the molecule concerned in a test-tube, we will get both isomers in roughly equal amounts. However, when molecules of this type react with other three-dimensional molecules there will be cases when one isomer will fit and the other will not, in the way that a left hand will not fit a right-hand glove.

Life-systems could either use both forms randomly or be selective. As it happens, the choice is highly specific, not random at all. Only one mirror image is used and the other has no biological activity. For example, vitamin C manufactured by chemists, is only half as effective as the same quantity of naturally produced vitamin from lemons, blackcurrants or spinach because the real stuff is 100 per cent active isomer, whereas the manufactured material is 50 per cent inactive. Further, and this is what is significant, not only all humans but all other animals that require vitamin C use the same isomer. Although there is nothing wrong with the other isomer, no such animal has yet emerged to use it. Randomness has failed us again.

This principle of shape is true of many other biochemicals. For example, there are eight 'essential amino-acids' which we cannot make for ourselves and must therefore obtain from the protein we eat. Each of the eight exists in the form of two isomers but only one of each, and always the same one, is used by all the animal species so far studied. No animal species has been thrown up to make use of the other isomers. On the face of it, an exceedingly large 'evolutionary niche' has been left unoccupied.

A similar situation exists in the essential polyunsaturated fatty acids. Like essential amino-acids, these are needed for reproduction and growth and have to be obtained from food. Linoleic acid is one of the essential fatty acids and is found in seed oils. The packets of certain soft margarines made from seed oils, have proudly informed us for a number of years that they contains cis-cis linoleic acid. What does this mean? It simply means that there are two forms or isomers of linoleic acid (cis and trans) and only one form can be used as an essential fatty acid. These isomers are not mirror images but just different shapes. The polyunsaturated fatty acid molecules have bends, which in the cis form are U-shaped, and in the trans form are in the shape of the steps of a staircase. Again, no animal is known that can use the trans form for essential functions.

The study of the way in which the chemicals of life behave in a test-tube sheds a further light on the question of randomness versus determinacy. Supposing we heat a mixture of amino-acids in a test-tube, is the product random? A simple question which can be tested experimentally. However random the mixture, the polymer that results is not random: the amino-acids will join together in certain sequences but not in others. The reason for this lies simply in the chemistry of shapes.

Any carbon-based molecule has a certain shape and when another molecule attempts to join up with it, the shapes have to fit. This means both the 'physical shape' of the molecules and also the 'electrical shape' and its strength, will determine whether two molecules fit or one fits better than another. There is a chemical and electrical specificity which determines what joins with what. In the words of Professor Fox: 'Our experiments in the thermal condensation of amino-acids of different types, and experiments in several other laboratories have demonstrated most strikingly that the amino-acids are ordered to a high degree in the resultant polymer' (Fox, 1974). This ordering is consistent with chemical determinacy even at this complex level. Chance takes a back seat.

Fox called the artificially produced polymers of amino-acids 'proteinoids', and they behave in a most interesting way. They will form themselves into tiny spheres and behave in some respects like living cells. In the right medium they will actually grow until they reach a certain size and then start 'budding'. These buds eventually break away to form new spheres which in turn grow and divide. These microspheres are producing new strings of amino-acids: in other words they are synthesising protein. Furthermore, they do so in predictable ways which run counter to the idea of randomness, and Fox himself comments:

> Darwinian selection operates on generated and evolved biosystems but not on those arising from a random (chemical) matrix. The studies of molecular evolution have revealed vividly that constraints in molecular possiblities have been dominant (Fox, 1974).

The reason why proteinoids can behave in this life-like way is that they can act as enzymes. Enzymes are active biochemicals that make more complex molecules, or break down existing molecules and order them to recombine in new forms. An enzyme is a protein that has, somewhere on the surface of its molecule, a zone into which a certain other molecule or combination of molecules will fit. Once a molecule has wandered into position, the electro-magnetic forces of the enzyme will either pull at it and split it apart, or clamp two molecules together to make a bigger one. When a number of amino-acids are joined together there is a strong chance that some part of the resulting protein will function as an enzyme. No 'vital force' or other mysterious power is needed to bring this about. It is simply a predictable result of the structure of amino-acid polymers. We can therefore assume that even before life appeared enzymes would have been at work producing biochemicals, just as they did in Fox's proteinoids.

In an enzyme, not only are the amino-acids arranged in a certain sequence, but the whole molecule must be folded in a certain way. When an enzyme called ribonuclease was synthesised in the laboratory using standard chemical techniques, the chain folded up as soon as the sequence of amino-acids was produced in the same way as it does in its natural form. Synthetic insulin similarly folds up into the shape of insulin made in the body. If such a sequence of amino-acids had occurred in the primordial soup, there can be absolutely no doubt that it would have done the same and possessed the same biological activity. If we want to make them ourselves, all we need to know is the sequence, then simply follow a set of rules.

While proteinoids can replicate themselves, a much more powerful form of self-replication is performed by DNA, which controls the production of protein in all living cells. The information in DNA is encoded in a sequence of simple 'bases', the smaller molecules of which compose its long double helix. There are four sorts of base and the code is provided by the sequence in which three of the different bases occur.

If we label them A, B, C and D, then two successive bases could be AB or CC or BD or any of the 16 possible combinations of the four letters. For three successive bases there are 64 possible combinations, for four 256, for ten bases over a million, and for 30 over a billion billion. The current evidence suggests that a group of three of the four letters provides the code for a single amino-acid. The words go to make up paragraphs which in reality describe (or code) a protein like insulin. Some sequences are meaningful and others have no apparent meaning at all, just as not all sequences of letters produce words. Even so, the number of possible 'words' is more than enough to contain the entire blueprint for the most complex organism.

Furthermore, DNA possesses a marvellous ability to transmit this code to future generations of cells with an almost infallible accuracy. There are two keys to this. One is the way in which the two strands of the molecule spiral round each other in a double helix. The other lies in the nature of the four bases from which it is built.

The four bases occur as two pairs; one member of each pair is an exact complement to the other in size, shape and electric charge. Hence, A will always combine with C and B with D. The two strands are wound round each other, and each base on one strand is paired with the opposite type of base in the corresponding place on the other strand. If the sequence ABCD occurs on one strand, then at the same point on the other strand there will be the opposite sequence CDAB.

During cell replication, the two strands uncoil and separate. When this happens each base on one strand attracts to itself the opposite kind of base, from those available in the medium around it, so that the sequence ABCD in the first strand would attract the bases CDAB, and the opposite strand made of CBAD would attract

ADCB. In this way each strand builds up on itself an exact replica of the one from which it uncoiled, and when both the original strands have done this there are two complete double helices, each identical with the original one. The molecule has replicated itself with absolute precision.

Fig. 6

Precise replication of an amino-acid four letter code in DNA

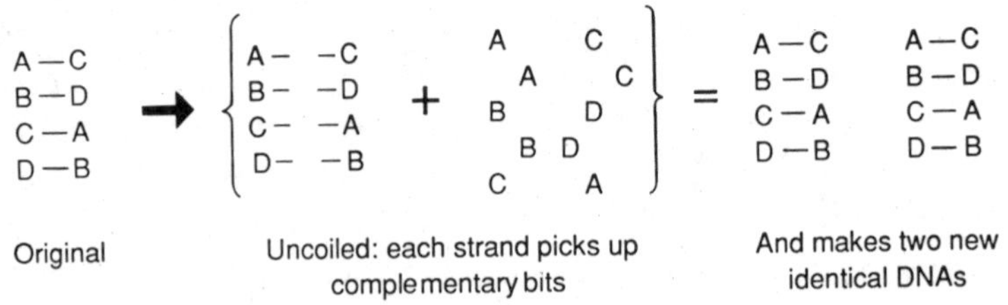

A—C
B—D
C—A
D—B

→ { A– –C, B– –D, C– –A, D– –B } + { A, C, A, C, B, D, B D, C, A } =

A—C
B—D
C—A
D—B

A—C
B—D
C—A
D—B

Original

Uncoiled: each strand picks up complementary bits

And makes two new identical DNAs

The bases are known as the purines and the pyrimidines. Since they were among the biochemicals made in test-tubes under model new-planet conditions, it is clear that they must have appeared in an abiotic environment in the primordial soup.

But how, you might ask, would the molecules coil and uncoil to replicate? Could this have happened in the primordial soup? Yes. You can unwind double helices in a test-tube simply by heating. The strands recoil when it cools. Indeed, this property is used routinely by molecular biologists to study coding sequences.

In the early days of the soup, proteinoids and DNA and many other molecules would have appeared. In shallow parts or in muds, the temperature would have been close to boiling in the daytime and cold at night. The daytime and nighttime temperatures would have been much further apart as the protective atmospheric blanket had not yet been woven. So it is possible that strands of DNA uncoiled during the day and recoiled at night. There would have been many different strands floating around and many different sequences. However even here the nature of the sequences would have been determined by thermodynamic considerations of the conditions and the catalytic medium to which adsorbant clays might well have contributed. As with the formation of the proteinoids, the sequences that were most stable would be the more common.

We do not know whether it was DNA or the proteinoids that became the first self-replicating structures. Very likely they began to act together from an early stage, as DNA and proteins do in an ordinary cell.

It may seem bizarre to claim that DNA and proteins simply got together to produce life. Yet we know this happens. Viruses are basically simple bits of DNA which slot into a cell's production lines and use its enzymes to selfishly replicate themselves regardless of the fact that they may be killing their host. The cell's RNA transcribes the information on the viral DNA, synthesises viral protein and new viruses. Hence we not only know from research that DNA and RNA can gain access to and join forces with foreign proteins but we actually know from our own experience every time we catch a cold. In which case an airborne virus hijacks the machinery of our own cells to replicate itself.

There is therefore absolutely no reason to argue that free-living DNA could not join with free-living proteinoids to produce a self-replicating system: hence life.

There is however a problem as to where the energy came from in the primordial cell. It may sound implausible to suggest that energy can be drawn into collaboration in such a simple way, but in fact it could have been just as simple as that!

The answer is that energy flows along the path of least resistance. If you boil a kettle and switch off the heat, it cools because its energy is flowing out of the kettle into the lower temperature environment. Lift up the kettle and feel the table underneath and it is hot, even though the kettle stands on legs: the energy has flowed from the kettle into the table. Had the bench been hotter than the kettle, the heat would have gone the other way: hence the electric cooker.

Now proteins require energy to reproduce – that is, they are *energy acceptors*. A donor and acceptor will gravitate towards each other and make sub-units, for similar electro-chemical reasons that opposite charges bring atoms together to make molecules.

So it would be quite natural for the energy-trapping molecules to use proteinoids or proteins to accept energy for them, which would, in turn, dissipate the energy by reproducing: the insertion of the DNA and RNA templates would then provide the degree of organisation that would lead to self-replicating life. The essential components of a primitive cell are not much more than energy receptors – proteins, DNA and RNA. Bacteria are even simpler as they do not have an energy receptor.

Once the happy union of DNA–energy–proteins began to co-ordinate their activities, something that was recognisable as living matter had appeared. With the appearance of those systems the corner from chemical to biological evolution had been turned. The corner had not been turned by the principles of the casino, it had been turned by the inevitability of chemistry.

There is, however, one powerful argument which has convinced many of the impossibility of transforming simple chemicals into life. It is that life is not simply the sum of its parts. The wheels and pistons, seats and steering wheel and all the parts of a motor-car can be piled into a box but they do not make a motor car. Professor Waddington's bricks do not amount to a habitable dwelling. Some-one has to join them in a special way to make a car work or to build a house. Similarly, although we are made of carbon, hydrogen, oxygen and other elements, we are quite obviously more than just a mass of carbon, hydrogen and oxygen. The answer lies in the laws of chemistry, and in chemistry the whole *never* equals the sum of its parts.

Take common salt: it is a compound of sodium and chlorine. Sodium reacts violently when placed in water; it raises the temperature to near boiling and behaves in a most unruly manner, dashing about, fizzing, bubbling and releasing hydrogen in an explosive way until it dissolves and turns into sodium hydroxide. Chlorine is a very nasty, lethal gas which was used to devastat-ing effect in the First World War. If dissolved in water it too behaves in an unsavoury manner, in time turning into hydrocho-loric acid – one of the strongest burning acids. Yet when sodium and chlorine are combined together as salt you happily sprinkle it on your soup. In chemistry, the total is never the same as the sum of its parts.

When life first emerged, it did not emerge on a barren, moon-like surface but in the rich field of opportunity presented by the primordial soup. It is easy to underestimate that richness. We have a tendency to think of organic matter as being more plentiful now than before life emerged, but, in fact, there was probably as much then as there is now, the only difference being that the organic chemicals were not then organised into life-systems. If we imagine the whole process being reversed and all living things – fish, insects, whales, grass, trees, and all the objects we have made from them such as food, clothes, furniture, houses – broken down and existing simply as chemicals lying on the ground, all the oceans dried up and their salts piled in a heap, then we may have an idea of the concentration of chemicals in that early time. This concentration would have varied from place to place – more so as water began to play its part. Just for that reason, many different combinations of chemicals must have existed in widely differing conditions. The oportunities for chemistry and life were in abundance.

In this way the great ladder was climbed. The elements forged in a star came together as compounds. When the temperature, atmosphere and chemical environment were right, those compounds based on carbon attained, step by step, new levels of complexity. At last they reached the level at which self-replicating compounds, such as proteinoids or DNA, were floating in an energy-rich environment and ultimately formed the first true biotic matter: the first living cell.

The instant that cell was formed the history of the future was written. The laws of chemistry which had built it were not suspended when it came into being. They continued to regulate and shape all its progeny from that day to this. Everything that followed, the leaves of a tree, the scales of a fish, the brontosaurus's enormous bulk, the bird's wing, the brain of man, all the amazing possiblities of life were contained in potential within the molecular structure of that single microscopic speck of matter floating in a warm, salt sea. Life is the property of chemicals and opportunity. If we started up the planet with the same conditions, we would logically expect life to occur again in much the same way.

True, we do not yet understand biochemistry with the same ease with which we understand the chemistry of salt. Our knowledge is new and fragmentary. The ability to construct whole molecules and to see them in three dimensions in a computer, gives us not only a new ability to operate in and understand the chemical dimension of life systems but also to defeat AIDS and any other mutant virus that the future holds in store. To future generations, the synthesis of a drug to defeat AIDS will seem as simple as the synthesis of, for example, urea is to us.

There is a continuum from the creation of matter to the creation of the stars, the planet and finally life. Like all other parts of that continuity, the origin of the planet and ourselves was governed by the laws of chemistry and physics: matter and chemicals simply responded logically to the different conditions.

4 The advent of oxygen

In life there is death. Since the beginning of records, people have believed in this fixed law of nature. But is there any good biological reason why our bodies should die? For the first species, the blue-green algae, had no such problem. Why, we might ask, has no animal species solved the problem? If the rules of chemistry determined the origin of life itself, then is the reason for death also buried in the rules of chemistry?

People have argued that death is necessary because without it, the planet would rapidly be overpopulated and the species become extinct. This argument does not really explain anything and it is quite possible to argue the opposite. People think we would overpopulate the planet because we like sex and keep on doing it. It is quite possible to imagine an intelligent species with less sexually focused interests and which only reproduced those who died by accident. Death is a puzzle.

At the start of life, the blue-green algae just went on budding for 1.6 billion years and then went into decline: but animals die. So 500–600 million years ago something – decidedly new – was involved with the origin of the animals. To understand what was new, it is helpful to know the chemical differences between them and the previous life forms.

Before attempting to construct an *a priori* role for chemistry in the origin of animals, we need first to answer a criticism that chemistry is irrelevant: that the time-scales are too short for chemistry to have created something so complex as life in the first place, never mind the even shorter span involved since the origin of the new dimension of animal life. In answering this question we will also be answering the first of Rattray Taylor's questions on the 'suddenness' of major changes, of which the change from chemicals to life was a major event.

The age of the planet can be estimated by calculating the rate of decay of the radio-isotopes in rocks and working out when they were created. The rocks giving the oldest dates have been found in the Morton Gneiss in Minnesota, the Baberton Mountain Land of Swaziland and the Warrawoona Group in Western Australia. These are about 3.5 billion years old. A pebble from a volcanic eruption in west Greenland has been dated at 3.8 billion years old. Moon and meteor rocks give dates of over 4 billion years. These give a more accurate estimate of the age of our planetary system since the moon cooled and its nuclear reactors came quickly to a halt. On this type of evidence, our planet is probably about 4.6 billion years old. However, a billion years must have first passed before the planet reached a temperature compatible with our type of life.

The first fossil remains of single-cell organisms resembling bacteria have been reported in rocks about 3 billion years old. There is some controversy as to whether these are fossils of living cells or just prebiotic organic matter. There are also some mysterious marks known as stromatolites in rocks that are 2.5 and 3.5 billion years old. Some researchers believe they were produced by the activity of living cells but they may have been forms of pre-life, the product of chemical rather than biological evolution. By 2.6 billion years ago fossils were being laid down of what were indisputable living cells: the first blue-green algae. It took another 1.6 billion years before animal life evolved and after that there was only the brief spell of 500 million years before *Homo sapiens* saw the light of day.

Some, including Hoyle, seriously doubt that this scenario offers enough time for chemical evolution to take place. It is argued that if it took 1.6 or more billion years for the blue-green algae to escape from their rut and evolve into animals, then chemical evolution would have had to occupy a much greater time span – i.e. many billion years. The time to produce life in the first place is seen as a far greater challenge than simply changing the nature of existing life into animals and plants.

In our view this approach is mistaken. The blue-green algae persisted so long, not because they were randomly changing until a new format was discovered but because the *conditions favouring them persisted*. The change happened when it did because of the cumulative action of billions of generations of the algae themselves producing oxygen.

Chemical evolution was not like that. During the cooling of the fireball, different reactions would have taken place under the different conditions. Each phase made its contribution to the subsequent evolution of chemicals and then life, which is simply more complex chemicals. The materials available and the conditions – chiefly the temperature and the concentrations of chemicals – were changing all the time.

As soon as the right materials and conditions came together the reaction would have been virtually instantaneous. If an amount of gas, equivalent to the weight of its molecule in grams (in the case of hydrogen this would be 2g), is reacting at a rate of 20,000 molecules every second, the reaction would take 20,000 years to reach completion! This piece of arithmetic tells us two things: first, reactions which do not seem to happen, or which we consider slow, may well happen if we think outside of the time-scales of our own lives. Secondly, the majority of chemical reactions which we actually know about must be occurring at an unbelievable speed. Most simple reactions occur immediately the chemicals meet. Miller produced his amino-acids in a week. Any chemistry student can produce a polymer in less than an afternoon, quite often without trying.

In the first cells to exist on our planet, the self-replicating DNA swam inside a mixture of proteins and other biochemicals. The latter provided energy and building units for the DNA, which in return directed the manufacture of more protein until the whole cell duplicated itself. It is known that bacteria and blue-green algae existed but it is almost certainly true that there were also viruses and bacteriophages (a virus which attacks bacteria) and many semi-life particles and chemicals which could hardly be expected to leave a trace. None of the first life systems developed nuclei and it was only at the end of their long, unchallenged monopoly – 1.6 billion uneventful years – that a new age of opportunity arrived. Then, in a short time, a whole series of radical changes took place and the boundaries of life exploded outwards. One of these changes was the coming of oxygen. Another was the appearance of cells with nuclei.

These more complex cells, like their successors today, had other tiny sub-cellular units, each performing a specialised function and each contained in its own membrane, held like packets in a sort of network inside the cells. It was this new inner organisation that made possible all the higher forms of life that followed.

The two kinds of cell are known as the prokaryotes (those without inner membranes) and the eukaryotes (those with them).

You and I and the lion (and the worms, the fishes, the trees, grasses and all other multicellular forms of life) are eukaryotic. In the eukaryotic organisms the different functions are organised into discrete geographical locations by subcellular membranes. In the prokaryotic organisms one function fuses with another in a rather sloppy mess. Effective and simple but with few organisational requirements. 'In fact this basic divergence in cellular structure which separates the bacteria and the blue/green algae from all other cellular organisms, probably represents the greatest single evolutionary discontinuity to be found in the present day living world' (Stanier, Adelberg, & Doudoroff, 1963).

The prokaryote cells unequivocally dominated the scenario and seem to have been alone throughout the Archaean period, from 3.4 billion years ago to roughly 2 billion years ago. Then fragmentary evidence suggesting some eukaryota begins to appear. However, the scene is still dominated by the blue-green algae and bacteria; it is not until about 800 to 600 million years ago that the metazoans (the beginnings of the animals) put in an appearance. At about 500 million years ago, the evidence of multicellular animal life explodes on the scene. When one thinks that the rest of evolution from the trilobites to the graptolites, jellyfish, shellfish, vertebrates, fishes, amphibia, reptiles, dinosaurs, birds, bees and the modern mammals, has been packed into this relatively short period, it is remarkable. How does one explain a sudden appearance of the blue-green algae, their long unchanging life-span and then the sudden appearance of multicellular life?

The most likely answer to the time-scale question is that the evolution of the first life from chemicals was a much simpler and hence faster game than people generally believe. It did not happen on a barren moon as people often suppose. It would have happened in a rich chemical environment based on fast chemical reactions. Chemistry only really slows down when the temperature drops to near absolute zero or when the more complex dimension of life chemistry – biochemistry – is reached. Once they evolved, the first life-forms, the algae, persisted so long because it took that time for the oxygen level to be raised sufficiently to permit the evolution of anything else.

At the beginning there was no free oxygen. Just about every element had combined with what oxygen there was – often violently. If you set light to a gas leak you may witness how dramatic this combination can be! With every element clamouring for oxygen during the fireball phase, there simply was not enough to go round. So a lot of elements and chemicals were left with their tongues hanging out – what chemists refer to as 'in a reduced state'. The job of the blue-green algae was to use the sun's energy to release some of this oxygen, from carbon dioxide, water and other oxides, into the atmosphere. But first, the oxygen they produced would have been taken up by those chemicals which didn't get enough oxygen in the fireball and were thirsting for it. Of those left, iron had a particularly strong thirst. The quenching of this thirst would have continued after the evolution of the algae and may have delayed the time at which oxygen tension could rise in the water and atmosphere.

P. Cloud, in an extensive study of this matter, considers that 'the oxygenation of the atmosphere was necessary to permit the change of life from fermentation of the anaerobic systems to oxidative respiration' (Cloud, 1976). This switch occurs in aerobic organisms which can also operate anaerobically at about 1 per cent of the present atmospheric level of oxygen. This is the so called 'Pasteur point'. Yeast, for example, can function both aerobically and anaerobically. Cloud has estimated that the origin of the 'eukaryotes coincides with an oxygen level having risen to about 3 per cent of the present atmospheric level'.

In summary, there is evidence of an immense period of constancy followed by a massive change towards air-using life, coinciding with the advent of oxygen. When the change came, it was fast and diverse, making possible the new, more complex eukaryote cells and the air-breathing animals. Darwin's concern that 'sudden generation was fatal to the theory of evolution through natural selection' is food for thought! The spread of these events is, however, consistent with the fast nature of chemical reactions.

There are some interesting ideas on how the more complex eukaryote cells came into being. Classically, one would suppose that the blue-green algae underwent a number of selective changes to produce inner organisation. Alternatively, specialised functions had emerged independently and simply joined forces in a symbiotic relationship rather like the way nucleic acids, energy-consuming proteinoids and energy-trapping molecules may have joined up to produce the prokaryotes.

This idea is not as bizarre as it first seems. The cell is made of discrete sub-units each with its own role. We have already seen this with sub-units joining forces in the form of virus infections.

Other sub-cellular units build up protein, store enzymes and provide the cell with energy. This question of energy is central to the whole existence of life, for any living thing must have a supply of energy if it is to work at all. We saw electrical energy used in Miller's experiment to create amino-acids, and at every stage energy in some form must be available to power the processes of life.

The basis of the energy cycle lies in the production of what is known as chemical energy. When certain atoms or molecules are excited by energy coming to them, they will group together to form compounds which hold or store that energy. For example, the cell adds a phosphate to adenosine diphosphate to produce adenosine triphosphate. Three phosphates can only be put together with a lot of energy. The three together hold the high energy in a slightly unstable molecule; when the third phosphate later breaks off, the energy is released. (The fireball must have produced lots of triphosphates – pyrophosphates and other energy sinks.) We feel this process in use when the chemical energy in the cells of our muscles is released as mechanical energy.

The first link in the chain is chlorophyll in plants, which is able to convert energy from sunlight into chemical energy. It was this process of 'photosynthetic' energy production that was used by the blue-green algae and it has remained the basis of all plant life ever since. In the prokaryotes there were simply molecules which soaked up the energy. In the modern plants, there are special sub-cellular units called *chloroplasts*, which contain these energy absorbing molecules.

The differences between plants and animals is that no animal can trap the sun's energy. Plants use this energy to build up pyrophosphates, then use them to make sugars, celluloses, proteins and fats. Animals must obtain their chemical energy from eating plants and breaking down plant sugars and fats. The only other way is to eat other animals who have in turn eaten plants or other animals.

Animals obtain energy by releasing that original plant-made energy into the body. In this process a vital part is played by another of the organised forms of the sub-cellular unit, the *mitochondria*. The mitochondria are able to make use of the oxygen brought to the cell from the lungs in the blood and to package its energy as high-energy molecules in a form the cell can use. They are therefore an indispensable part of any oxygen-using system.*

A fascinating theory about how cells came to possess organelles through symbiosis was first put forward by A. P. Finnan in about 1870. Some 100 years later Lynn Margulis updated the story in several papers and in her book *Symbiosis in Cell Evolution*, published in 1981. The idea of symbiosis is all the more interesting to us because it is similar in principle to our attempt to construct *a priori* the origin of life. Margulis considers the classical view that the more highly organised forms of life evolved from blue-green algae or bacteria 'by the accumulation of selectively advantageous mutations' and finds it 'inconsistent with many facts', including the problem of the discontinuities.

It is known that, given the right conditions, mitochondria can live as independent organisms, and Margulis suggests that, at the beginning, they already existed alongside the blue-green algae, floating around as separate units in the primeval seas. What happened next was that they migrated into the interiors of the cells of the new eukaryotes to be, and lived there in symbiosis with the cells, receiving nourishment from them and, in turn, giving away their energy in an extension of the argument we presented previously on how energy and protein synthesis were co-ordinated. In support of this theory is the fact that mitochondria do not share the host cells' DNA: they have DNA of their own which they pass on to the new generation through the mitochondria contained in the mother's ovum (the male sperm has no mitochondria). Genetically, they lead a separate life from their hosts.

There are two good chemical reasons for such a symbiosis. First the production of high-energy molecules with no use would be like a factory producing electrical batteries with no sales. The workers would lose interest and stop production. This analogy with the

* Mitochondria are independent particles living inside cells; they operate the Krebs citric acid cycle which combines oxygen and hydrogen for energy production. Because its DNA is reproduced on its own, mutations are specific and not due to male/female interaction. Mitochondrial DNA has therefore been studied to seek information on our relationships with chimpanzees and gorillas. Rates of mutation have been calculated. It mutates 10 times faster than the cellular DNA: it has 16,000 base units. 20 out of 1000 bases change in a million years, 2000 in 100 million years and therefore 10,000 (i.e. the large part of it) since the beginning of animal life. Yet surprisingly, the Krebs citric acid cycle and the basic mechanism for energy production is the same in different species. Even at the level of coding for protein, cytochrome C (a key energy-converting enzyme) sequences have been studied by Margulis who found few and more often only one codon difference between species.

biochemical process is fairly accurate. Firstly, mitochondria feeding a system with energy would be far more active than those that were not. Secondly, it has recently been discovered that mitochondria 'import' from the cell certain building units for their own membranes: these are 'lipids' or, more precisely, 'phospholipids' about which we shall hear more later. Hence, the cell can assist in the formation of the mitochondria's own protective membranes. The cell both gave to and received from the mitochondria components important to the function of both.

In a similar way, on Margulis's theory, the chloroplasts which package the chlorophyll inside the cells of vegetable life could have existed independently and migrated into the cells of eukaryotic plants. Whether or not this theory is correct, it was the mitochondria and chloroplasts contained inside the eukaryotic cells that were the secrets to success of the new order of life. This key event in cell design combined with another even more important change occurring around the same time, to change the living world out of all recognition: the advent of oxygen and the air-breathing systems.

It is not surprising that events at this critical point in evolution are somewhat confused. Considering that all our knowledge comes from the fossil record left behind hundreds of millions of years ago by creatures of microscopic size, it is surprising that we know as much as we do. One thing is clear, however: the dramatic change that took place when the algae produced the right amount of oxygen led to the emergence of air-breathing, multicellular animal systems. Once again the change in chemistry is undeniable, and it does not seem that it was a struggle for survival which brought the animals into existence. There is no reason to suppose that there was not a plentiful supply of all the necessary nutrients, including high-energy compounds, built by the algae and waiting to be used by the new creatures that entered this virgin world. It is more likely that wealth, stimulation and co-operation, not bitter competition for meagre resources, was the great originator.

It is impossible, on the basis of our slender knowledge, to establish anything like a chronology for this transitional period, let alone a chain of cause and effect. All we can do is point to the four great changes that occurred and surmise that they are very unlikely to have been unrelated. These four changes were (i) the advent of oxygen; (ii) the arrival of the eukaryotes; (iii) the development of multicellular life systems; and (iv) the division of living things into the two great kingdoms of plants and animals.*

The first change held the key to the future. The arrival of free oxygen created a new set of conditions as important as were the new conditions created by the chemicals associated with life's origin at the very beginning. The last was the most dramatic of the four innovations, but in one sense it was the least revolutionary, for it came about simply through the marrying-up of the other three. A plant is a eukaryotic, multicellular version of the forms of life that existed before this great period of change. They were nourished by sunlight and the chemicals of the earth. An animal is a eukaryotic, multicellular, oxygen-using organism nourished by the plants and bacteria.

While we can only surmise on the symbiotic or other mechanisms whereby the development of cell membranes led to co-operating intracellular packages and then to the multicellular organisms – we merely observe that there are no known multicellular prokaryotes – we do, however, have a very clear *a priori* reason why oxygen made possible the emergence of more advanced forms of life.

The oxidative breakdown of plant products to produce energy is more efficient than the techniques of obtaining energy without oxygen. Systems like bacteria and yeasts can work without oxygen to break down long molecules into bits with two or three carbons (e.g. alcohol) but no further. Oxygen-using systems can burn these bits and so obtain much more energy – eight times as much, in fact. The chemical process by which they do this is the Krebs citric acid cycle which is universal in all oxygen-using animals and occurs only inside the mitochondria.

This high efficiency in energy generation presented the first oxygen-users with a new opportunity and a problem. They produced more energy in unit time and could therefore grow faster than the algae on which they fed. This meant that the food in their neighbourhood would soon be eaten and they would be forced to adopt one of two strategies. The first was to go out and get their food by moving around – the strategy later adopted by animals that learned to drift like jellyfish, swim like the shark, or run like the

* Life is never as simple as in books, but to avoid complicating the argument unnecessarily we have omitted such ambiguous organisms as fungi and those single-cell protista which sometimes obtain their energy from photosynthesis and at other times swim around and digest food particles. Interesting as they are they do not affect the thrust of our argument.

cheetah. The second was to sit still, suck in the water and filter out the food, as oysters do. We can see here the dominance of nutrition in evolution if we consider for a moment what is ultimately the root of the difference between the cheetah and the oyster: it is a matter of how they get hold of their food. The fact that evolution concerned different ways of getting food is such an obvious statement and the business of eating is of such a commonplace nature, that its central importance has been overlooked.

The difference between plants and animals was lessened, though certainly not eliminated, by the fact that when multicellular plants arrived they made some use of oxygen and thus increased their efficiency. During the day plants take in carbon dioxide from the air, use it in conjunction with water to build up compounds by photosynthesis, and give out oxygen. At night, however, they take in oxygen and for a few hours become aerobic organisms. They can therefore reproduce and grow faster than they could do otherwise, but still nothing like fast enough to keep up with the animals that live off them. Unless there is some check on population, animals will always end up outstripping their food supply, as Malthus observed about human communities. Hence, the death of an individual is not a solution to the problem although the death of an entire species would be. We believe this actually happened on several occasions.

The different phyla were not competing with each other but were simply exploring different ways of getting food. The amoeba's technique of folding itself around food developed into the forming of a canal, from mouth to anus; food passed through this canal and the nutrients were extracted during transit. Some animals moved themselves to chase food; others simply sat still and moved the food into themselves. It was, from either viewpoint, *movement* which provided one of the keys to the new life-forms.

Can we stretch our imagination to speculate on how 'movement' evolved? Muscles can only contract. No pushing muscles evolved. The contraction is caused by an electrical pulse with the energy provided by the pyrophosphates. The advent of oxygen would have forced whatever it was that was about to become an animal to make high-energy triphosphate in large amounts. As energy flows from high to lower energy states, it would be natural for the high-energy molecules to release their energy as soon as they bumped into a suitable acceptor.

Some of this energy would be given to protein enzymes which made big molecules out of small ones. Other high-energy molecules would come into contact with a totally useless (until now) protein with no enzyme function. The energy-burst made it grimace – it doubled up in pain – it contracted. The broken high-energy molecules then went back to be refuelled. The contraction would make this useless protein move. A rudimentary muscle was emerging.

Protein that twitched would have had to find room to twitch. Room would most readily be found at the periphery. In that position the twitching would mainly cause movement where there was least resistance, i.e. in the cell surface, and so the cell itself would have made its first faltering movements through the water.

Co-ordination would have followed. Cell membranes possess many chemical substances similar to the chemicals in nerve fibre chemistry. As soon as electrical charges were produced as a by-product of the surplus energy it is most likely they would discharge around the membrane and introduce 'order' into the twitching, contractile proteins. Lined up on the outside they were sparked off one after the other in response to the flow of electrical current. Animals had learnt to swim. And it all happened because a protein was capable of dissipating energy mechanically instead of chemically.

The above description may or may not be what happened but it does show how movement can be explained in very simple *a priori* terms, invoking concepts of no greater complexity than the simple production and release of energy. Movement was a response to the new, highly efficient production of energy created by the advent of oxygen. The energy had to be dissipated in the same way that water has to run downhill and movement was an important way of dissipating it.

In the case of the single-celled paramecium, the surface is covered in tiny hairs which 'row' in an astonishingly co-ordinated manner as the signal flows from one muscle to the next. This very simple computer programming algorithm for an electrical charge to contract a fibre was to be developed, with many additional subroutines, into the use of fins, wings and legs, and the mouth of an oyster. The origin of movement was possibly just a by-product of the production of abundant high-energy molecules from food and oxygen. However one looks at it, the food and oxygen had to come before the movement.

* * *

We can see how oxygen was a gift which created the animal and how energy production could have created movement. There were many other exciting innovations specifically associated with the arrival of animals. One of these was the new concept of death.

The 'accepted view' would say that animals die after they have reproduced: indeed some die in the act of doing it. The female praying mantis tears off the head of her mate as he is in the process. Her expression of gratitude is repaid by a sudden burst in the flow of sperm from the decapitated. The male black widow spider is also disposed of by the satisfied female and the poor male bee is disembowelled as a reward for giving his all to his queen. Therefore, it is argued, there was no selection pressure to solve the problem of death. But why have death at all when it did not exist in life before animals?

To try to answer this question, we need to think more about what was involved in becoming a multicellular animal. There are two important differences between legs and the paramecium hairs. First, the bigger the animal system the greater the distance from the mouth–anus canal to the moving parts. To reach these distant parts, the nutrients and oxygen had to be transferred from the canal or exterior gills by a vascular system. Secondly, a nervous system for control became an essential feature. It is no use having two legs if they walk in opposite directions. Hence we can add to oxygen the nervous and vascular systems as key new dimensions of multicellular animals which had not existed before.

In early systems and most of the animal phyla, the nervous system is a network. In the vertebrates it has a focal point: a brain. Animals graduated from networking IBM PCs to the genuine, multi-tasking, central computer, running many PCs and terminals.

Indeed, it seems as though the nervous system controlled everything and actually ran animal evolution: it could be argued that everything else was secondary. The evidence for the early dictation of events by the nervous system is the way in which it has a primary control over almost every body function. The central role of the nervous system is so obvious that it is seldom considered. One of the first systems to develop in the human embryo is the neural tube out of which grows the spinal cord and the brain. The primitive spinal cord extends into arms, legs and all other distant parts of the body, whilst at the head, the brain develops into eyes, ears, nose and tongue. The neural tube is the earliest and most central feature around which the embryonic development weaves its magic. But it is paralleled by the development of the heart which pumps blood and nourishment to the developing tube. We will return later to this evolutionary relationship between the vascular system and the brain.

If the nervous system was a key to the origin of animals, then, following the logic of movement evolving out of energy dissipation, the brain and nervous system would similarly have to be a primary acceptor of energy.

This turns out to be the case. The priority given to the nervous system as an energy user is absolute. In starvation, body reserves are devoted to its maintenance while other tissues are sacrificed. Its measure is seen clearly in the newborn baby. By studying babies born without any brains (anencephalic) compared to those with normal brains, it has been found that a staggering 60 to 70 per cent of the newborn's energy from food and reserves is used by the brain for growth and maintenance. Indeed, the brain is so rich in lipid that its metabolic rate per unit of cell mass is phenomenal.

Here is a clue to the choice of priorities for human development. Even the human adult brain uses as much as 20 per cent of the body's energy although it occupies only 2 per cent of the body. (It is a fascinating sidelight that the human intestine which deals with our food, although reporting to and under the eye of the 'Big Brain', has such a complex network of regulating nerve fibres acting without supervision, that Peter Milla of the Institute of Child Health in London refers to this network as the 'Little Brain'.) There was, however, a difference between the high-energy status which produced muscle movement and that which produced nerve function: energy was dissipated through *mechanical* energy in muscle but *electrical* action in the nerves. The electrical function can never stop; if it does, the brain is dead.

This last point tells us of a parallel development between the brain and the blood supply. Following in the tracks of the nervous system is the vascular system which carries oxygen and nutrients and removes waste products. The two systems evolved together with the brain being critically dependent on its blood supply; clamp off its supply for five minutes, and the brain dies. We can live without food for 30 days and, in a temperate climate, without water for five, but after no more than five minutes without oxygen the brain is dead. The blood vesels supply the brain and the tips of its peripheral nerves: hence the parallel between nervous systems and blood vessels. Clearly, the brain is the most crucial organ to the genesis of an animal and to death.

Was anything else new in the chemistry of the nervous and vascular systems which might help explain death? In fact, there was. The chemistry of the nervous and vascular systems is dependent on a group of chemicals which had not featured before the advent of multicellular animals. Sixty per cent of the structural or building materials of the brain and nervous system is, surprisingly, fat. Not any kind of fat: but lipid or 'structural' fat.

Structural considerations for prokaryote cells were of minor importance. Hence a new biochemistry came into play alongside the evolution of the multicellular animals. Protein and DNA chemistry had been the name of the game for the bacteria and blue-green algae. The higher organisation of multicellular systems introduced the chemistry of membranes. This new chemistry was the chemistry of lipids.

Special fats are involved in the membranes and these include polyunsaturated fats. Saturated fats are hard fats, such as those used to make candles, whereas polyunsaturated fats are liquid and form oils. The brain sends and receives many millions of messages every second and is the most membrane-rich system in animals. Blood vessels are also highly dependent on membrane fluidity. The blood vessels of modern animals use polyunsaturated fats to provide the flexibility that allows them to absorb the pressure waves generated by the pumping heart. Without that flexibility we get hardening of the arteries and a fatally inefficient vascular system.

Interestingly, the brain and vascular system of modern animals depend not just on any polyunsaturated fat, but on the most polyunsaturated types. The more active the functions, the more use is made of them. Sixty per cent of the structural material of the photoreceptor which turns a photon of light into a message for the brain is the most polyunsaturated fatty acid: docosahexaenoic acid which has 22 carbons and 6 double bonds. This structure sends, from one plane of vision only, 10,000 signals approximately every $\frac{1}{25}$ of a second to the corresponding network of brain cells which is then integrated by higher brain cell networks for interpretation. Modern office computers may have a 16 or 32 bit bus carrying the signals. Just one plane of vision is your eye is served by 10,000 bit bus!

The important downside of this equation is that the polyunsaturated fatty acids are highly susceptible to oxidation. The specialisations of both the vascular and nervous systems depend on membranes made with lipids that are readily destroyed by oxygen: yet the brain is, paradoxically, most critically dependent on both oxygen and oxygen-vulnerable building units.

One of the reasons why we associate fresh air with health is that bacteria, having evolved in an oxygen-free environment, die if they are exposed to enough of it. In fact free oxygen will damage or kill any cell if it penetrates its protective coat. One reason for this is that it can destroy the enzymes on which all cells depend by attacking a key chemical group known as sulphydril. This active group is a hydrogen attached to a sulphur which itself is part of an amino-acid in the protein daisy-chain. What the oxygen does is to steal the hydrogen atom, with which it readily reacts. This leaves unattached sulphur groups which respond by joining up to form 'sulphydril bridges'. The effect is to clamp the two chains of the protein together in a rigid and unresponsive shape and seize up the machinery: the protein affected can no longer function. Biochemists who extract and study the working parts from inside the cell know how vital it is to protect them from air or oxygen. Like the bacteria, the inside of the cell has a chemistry of a kind that predates free oxygen.

There are a number of other substances within the body which oxygen destroys; among the most important are the polyunsaturated fats and fat-soluble vitamins A and E. They may have played only a small part in the structure of the blue-green algae. In modern animals the polyunsaturates are employed as essential building units for membranes. It was these types of membranes which seemingly enabled the eukaryotes to create the sophisticated sub-cellular organisation that led to higher functions such as signal transmission, vision and the other senses. In fact, this new biological chemistry of oil-soluble nutrients actually made possible the evolution of the brain and the nervous and vascular systems upon which the multicellular animals depended. Yet, astonishingly, they are extremely vulnerable to oxidation.

One would have thought that evolution would have used oxygen-resistant components for such key materials in the oxygen-using systems. But it did not. Perhaps there is a clue here to the question of death.

Here then is the paradox: the evolution of multicellular aerobic organisms was wholly dependent on an element which was highly toxic to them. The answer is ingenious: the oxygen is packaged. Within our bloodstreams are red cells and inside these are molecules of haemoglobin, which takes up the oxygen from the air in our lungs and releases it where it is needed in a manageable form.

Another protective measure that benefited the aerobic organisms is a range of 'anti-oxidants'. These are biochemicals which are more susceptible to oxidation than the cell constituents they are protecting, and which take up any free oxygen so sacrificing themselves. At least two vitamins do this job: one of them, C, is soluble in water and the other, E, in fats. The necessity of these anti-oxidants was tragically illustrated by what happened to a number of premature babies placed in oxygen tents immediately they were born. The oxygen destroyed the retinas in their eyes and they went blind. It is thought that because they had been born too early the mother had not been able to transfer enough of the protective nutrients to the child.

The oxygen-sensitive polyunsaturated fats so necessary for the construction of the brain and vascular system cannot be made in the body. They occur at the base of the food chain in plants. The odd fact is that they occur in the same place as vitamin E and C which protect against oxygen attack.

As early as 1958, H. Dam and co-workers showed that 'crazy chick' disease, which killed large flocks of chickens, was due to old feeds which contained polyunsaturated fats but were deficient in vitamin E. In old feeds the vitamin E had been lost, destroyed by oxygen while protecting the polyunsaturates. Loss of the protective agent in the feed resulted in destruction of the cerebellum in the chicks (Dam *et al.*, 1958). Recently at the Nuffield Laboratory Professor Pierre Budowski showed that the period during which the chickens died was the same period in which the cerebellum was growing rapidly and accumulating its polyunsaturated fatty acids. The remarkable conclusion is that had nature not made vitamin E alongside its polyunsaturated fatty acids, advanced nervous systems could not have evolved. Today, Hoffman la Roche makes lots of vitamin E to protect chicken brains.

It was no accident or random event that located the oxygen-sensitive nutrients and the protective agents together. The chemistry of the blue-green algae and the plants arose in an oxygen-free condition. Many of the chemicals used were the sort which thirst for oxygen, so the coincident appearance of oxygen-sensitive chemicals is not surprising. When oxygen-using life emerged with nervous and vascular systems dependent on oxygen-sensitive materials, they had a problem. Chemicals like sulphydril groups and polyunsaturated fats employed in function and structure were under threat. So chemicals which had an even greater thirst for oxygen and which occurred in food alongside these materials suddenly took a new significance: self-sacrifice. There is a mechanism for bringing some of these agents back to life but in the long run, they became Kamikaze protective agents.

The realisation that emerges is this: the design that was to become essential for the brain and for the nervous and vascular systems was being drawn even at the early moments when there was no oxygen in the atmosphere over 1.6 billion years before air-breathing species emerged.

With a conservatism that we are coming to expect, oxygen-using systems retained the basic intracellular processes of the previous epoch, with add-on bits reacting with oxygen. It is also significant that the protective elements themselves have remained proof against random variation. Vitamin E is universally used, as is vitamin C. Haemoglobin is the material that all animals use to transport oxygen through their bloodstreams, and the differences between species are minor. Even the crab, which has not red but greenish blood, is not as different as it seems. The normal haemoglobin molecule is based on an atom of iron – the crab's is based on copper, but is in other ways remarkably similar. Still more remarkable is the chemical similarity between haemoglobin and chlorophyll in plants, the molecule of which is based on magnesium, but is otherwise almost identical to haemoglobin.

Yet, despite the success of these protective measures the multicellular organisms still faced the inevitability of death. A single-cell organism does not die. When it divides in two neither part has any more claim to be the parent than the other, and the same is true of later generations containing much larger numbers of its descendants. Even if many of them are killed off, the first parent continues to exist equally in all the others. Many of the early algae and bacteria did of course die and perhaps were killed by malignant bits of DNA or RNA. But their progenitors continued to exist in the survivors. The same is not true of a multicellular organism. When its time comes to die it cannot console itself with the thought that it will live on in its children. That is only metaphorically true. The simple reality for that organism is death. For single cells, death is not inevitable. Hence we have two features linked to the introduction of death – the use of oxygen and multicellular organisation: the latter being dependent on the use of new, oxygen-sensitive lipids.

There are currently two popular theories of ageing. Professor Darmandy has described this process as 'biological rancidification' which is, effectively, progressive and irreversible oxidation. Others argue that ageing and death are programmed, a form of cell commitment. Bearing in mind that external influences are likely to play a part in cell commitment, the two ideas are not incompatible. There is also growing evidence that cancer may be induced by oxidative damage to DNA.

Although we are doomed to die, some organisms are much more successful than others at delaying the inevitable. For a start, the most successful ones are vegetables. One of the longest-lived organisms in the world is the California redwood; some of these giant trees are known to be more than 1000 years old. Even that age may have been exceeded by individual trees of other species. On the Greek island of Kos there is a plane tree under which Hippocrates is said to have taught in 400BC. There is no doubt that the tree, highly protected and propped up in places, is of great antiquity and it is possible that the story is true. There are also ancient trees in Asia under which great masters are said to have taught thousands of years ago: the trees are cared for by priests and protected by temples built round them. In more temperate climates oak trees can live to great ages and their lives usually end when they are cut down for timber rather than when they reach their programmed biological end. The oldest accurately dated tree is a Bristlecone pine from the White Mountains of California. It was found to be 4600 years old by ring counting.

No animal can match this sort of life-span. The strongest competitor is the giant tortoise which may live for 200 years. Other reptiles also live to great ages, perhaps because they have low temperatures and all their chemical reactions take place at a slow pace; but even the longest-lived of them cannot begin to match the trees.

The explanation for the long drawn out period of 1.6 billion years of blue-green algae with a total absence of animal evolution could be stark simplicity. The animal phyla could not evolve until there was oxygen, whereas the blue-green algae had no need of it, and simply went on budding and reproducing until they themselves had produced enough oxygen to substantially alter the conditions on earth. At this point the oxygen-dependent animal phyla could come into being.

But the design of animal systems contains a self-contradiction. Their complicated systems need oxygen, but oxygen in its free state acts on them as a deadly poison. Was this contradiction a self-destruct button?

5 The colonisation of land and air

By the end of the first great era of oxygen and opportunity, the scope of life had expanded in almost every way up to the boundaries that still contain it today. Instead of simple self-replicating molecules, one-celled anaerobic algae and bacteria, there were multicellular plants, fungi and animals breathing oxygen and using the same basic chemical processes that have remained in use ever since. The major categories of life systems were already in place and at that level they have never been added to. Their home was the sea. One great frontier remained, however. Up to this point life existed only on sunlit rocks, in mud or water into which the salts from the land and its volcanoes were being washed by the rain. The land and the freshwater rivers remained sterile and unoccupied.

The first living things to cross this frontier were the plants. In the vascular plants a structure called the xylem carries the water and the phloem carries the nutrients. In water plants the phloem and xylem occupy the centre of the stem. In land plants, however, the phloem and zylem are on the outside of the stem, supported by a rigid structure in the middle.

For the plants, this transition was comparatively easy because in one sense all plants remain water organisms. With the exception of some desert plants their roots must always be in a medium that contains some water, however little, and there is only a difference of degree between a seaweed living entirely surrounded by water and a semi-desert acacia drawing its supply droplet by droplet from the deep water tables.

For the plants that moved to fresh water the move was even simpler. Sunlight, water and carbon dioxide are the major requirements of any plant, and they can be found in fresh water as easily as in salt. It is likely that at the beginning the freshwater regions possessed a higher nutrient content than they do today. More importantly, the shallows and land regions were beckoning the plants with their offer of plentiful supplies of chemical fertilisers left over from the original fireball. The colossal sizes to which the vegetation grew once it took root testifies to the greater richness of the land at that time compared with today: it was an age in which giants flourished.

Long before the pioneer animals moved hesitantly out of the sea, it is likely that the land and its freshwater areas were richly furnished with plants of many species – a wonderful and untapped source of nutrients of all sorts for any animal that could make the land its home. When land animals did evolve they were presented with a new age of opportunity – a whole world full of rich diversity of habitats and foods, all suited to animals of different kinds. It is no wonder that such a spectrum of contrasting species arose as the phyla fanned out through new found strata of wealth. Once more, there was no need for competition. It was not a harrying struggle for survival that moulded the new shapes of life but rather the sheer wealth of possibilities. If ever there was an Eden, those first land animals entered it.

While accepting selection as a factor, Pirie is more reserved about *when* it comes into play:

> The essence of Natural Selection is that the amount of matter organised in the form of successful systems should increase at the expense of that organised in the form of the less successful. It can therefore operate only when there is a shortage of something. (Pirie, 1972)

To get to the land, the animals had to surmount difficulties that had never confronted the plants whose roots still remained comfortingly in water. What sea offers, and land and fresh water do not, is constancy. The temperature of the sea off Long Island is little different from Land's End. The temperature of mid-Atlantic water between these two points shows but small variations from summer to winter and from day to night. There are big differences if we compare the Arctic seas, which are only just above freezing point, with warm, coral lagoons at the equator: but those extremes are separated by thousands of miles. There are plenty of places where a land animal can experience even greater contrasts of temperature between day and night without moving from one single spot.

Another important constant of the marine environment is its salinity. Again, the salt content of the water off Land's End and Long Island is very similar, and even between the Arctic and the equator the differences are very small. Figures given by the *Biological Data Book* for the salinity of the Arctic Ocean are 32–33g per kilogram of sea water compared to 34–34.6 for the Antarctic and 34.5 for the North Pacific.

In one way this is deceptive. Although the salinity does not vary geographically it has varied a great deal over time. The earliest pools of water on the planet's surface must have been extremely thick, hot, super-concentrated solutions of many salts; since then the salinity has been much diluted as more water vapour condensed into rain. The proportions of different salts have also changed greatly, and in particular the balance between sodium and potassium salts has been reversed. At the present time the sea contains about 30 parts of sodium to every part of potassium.

It is not known for sure why the proportion of salts in the sea changed as much as it did, but there are two possible explanations. One is that sodium, being the more reactive element, was involved to a greater extent in the formation of rocks; it was then less likely that the sodium in the rocks would dissolve in water. The other is that potassium was selectively removed from the water by being incorporated into the cell systems living in the sea. Every gardener or farmer knows that his plants need potassium. The leaves of watercress contain 3.14g of potassium and 0.6g of sodium per 100g of dry plant tissue. The broad bean leaf has 22 times as much potassium as sodium and cocksfoot grass 10 times as much. Plants take potassium out of the soil and if it is not returned in the form of compost or fertiliser its concentration falls to the point where the plants are enfeebled and eventually stop growing altogether. However, a lot of plant material would be needed to explain such a dramatic shift in balance.

Either of these processes, or a combination of both, could account for the reversal of the salt balance of the sea. Whatever the cause, it is clear that the change took place and the response of life systems to it has highly interesting implications for evolution theory. In a sense their response has been not to respond. They have gone on maintaining an interior environment that closely mimics the exterior environment from which they first evolved.

If we look inside the plant cell we find that the ratio of potassium to sodium salts is astonishingly similar for plants of all different kinds, both marine and land species. It seems that for billions of years living cells have made no radical change in this respect. In the same way in which they have stuck to their oxygen-sensitive chemicals, all have clung to the salt conditions in which their first ancestors emerged: and, when they could no longer find them, they recreated those conditions inside themselves.*

In an even more remarkable way, the same thing is true if we take multicellular systems and examine their interior 'seas' – for that is in effect what the blood plasma in animals or sap in plants provides. No active cell can live in dry surroundings: it must always have moisture from which it can absorb the salts and other nutrients it requires, and this is just as true of the cells in a multicellular system. That system must maintain an internal environment which provides for the chemistry of its cells, and whilst the organism as a whole may live on land, its constituent cells each live as cells have always lived: in water. Our own tissues, like those of other animals, are about 80 per cent water. But what of the salts in the water? By the time the animals evolved, there was far more sodium in the water than potassium.

The fluids circulating through the bodies of animals, not inside the cells but around them, are known as plasma, and if we analyse the plasma in different animals we find that unlike the inside of the cell there is more sodium than potassium. For the rat the figure is 35.6 and the human 34.1 times more sodium: much the same.

In plasma, sodium has displaced potassium to become by far the commoner element. Now this is just what we would expect to have happened from the history of the sea. Multicellular life did not arrive until about two billion years or so after the first single cells – over three-quarters of the way down to our own time. By then much of the change in the sea's salt ratio had occurred.

Is this simply a matter of stubborn conservatism – an impoverished aristocrat refusing to give up the style of life he finds it ever more difficult to sustain? Something like that is the usual explanation: first the cells and then the animals' blood systems became shut off from the sea and so went off on a self-contained course which did not change. Is there not more to it than that? Professor Baldwin of University College London had a pertinent comment in his

* As mentioned above, it is calculated that mutations occur in the cells' genetic DNA at a rate of 2 per thousand bases every million years. In DNA with billions of bases, this offers the potential for a significant, random change. The question is whether or not random change is meaningful. Life has clung to its basic format for somewhere between 2 and 3 billion years. With such a large potential for change and a lack of it in practice, something must hold life systems in a consistent direction.

delightful book *An Introduction to Comparative Biochemistry*, where he echoes the words of Claude Bernard on the constancy of the 'milieu interieur' written a century before:

> Thus it is possible that at some time in the past the composition of the sea was such that life was possible in it . . . instead of being surprised that the bloods of different animals resemble each other so closely, we must realise that it would not have been otherwise. The composition of the blood has remained the same because the conditions under which life is possible remained the same (Baldwin, 1949).

In other words, those were the conditions in which chemistry could produce our form of life. There is a limited number of reactions on which life can be based and those reactions take place within a narrow range of catalysts, temperatures and salinity of the right sort. When those conditions arrive, the new form of life will develop rapidly: until they do, life cannot develop at all.

There is further evidence for this in the salt levels in animal tissue, as well as in plasma. As Baldwin, and before him Haldane had pointed out, salt levels are somewhat similar for widely different species, even for marine and land animals. For example, in muscles of herring, cod, and salmon, the atomic ratios of potassium to sodium are 3·0:1, 2·4:1 and 1·8:1 respectively compared to chicken, lamb and beef with ratios of 2·3:1, 2·1:1 and 1·8:1. Indeed, concentrations of sodium, calcium, magnesium, cholorine and phosphorus are more remarkable for their similarity than for any differences from species to species. The exceptions are rare and such exceptions which do exist (the dog for one) deserve study as to how they came about and why they are rare and not more frequent. Red cells have much higher potassium to sodium ratios and are much closer to plants in this respect than muscle. The haemoglobin of the red cell is so close in structure to the chlorophyll of the plant chloroplast, it makes one wonder if the genetic information for the red cell came from a mutation of the chloroplast data.

The responsibility for keeping the right salt balance in the organism as a whole falls on the kidney. To maintain the correct levels it will excrete more or less sodium or potassium into the urine, and the accuracy with which it performs this task is remarkable. Normally the concentration of potassium in human plasma is about 4.5 units per litre. If you eat a meal of potassium-rich foods such as lean meat, vegetables and fruit, the concentration of potassium in your plasma will hardly rise above 4.7 units per litre but the concentration in your urine may well rise twenty- or thirty-fold. This is just as well. If the level in your plasma went above 5.7 you would be in serious danger of cardiac arrest. The beating of the heart is regulated by a series of nervous impulses caused by a movement of charged particles such as potassium, sodium and calcium ions,* and this mechanism is upset by changes in the salt balance.

Interestingly, the control of kidney function can be traced back to the salt- and water-sensitive regions in the brain. The kidney works at maintaining constant levels of salts, including calcium and phosphorus, with an efficiency that leaves us quite unconscious of it – until, that is, we stretch the system beyond its limits. Baldwin gives the example of miners and stokers who work for long periods at high temperatures. They sweat profusely and so lose large quantities of water and essential salts. If they then drink a lot of water they can suffer from violent and painful muscular cramps, which can be avoided if they take enough salt with their drink. The reason for this has to do with the chief protein in muscle, myosin. This protein is insoluble in pure water, or in water which has a high concentration of salt. It can only be dissolved – and therefore only function – in water with a level of salt held between quite tight limits.

Not all animals, of course, developed such effective systems to control their salt balances. While all marine forms have coped over the millennia with a changing sea, they have on the whole done no more than that. Most of them die within hours or even minutes if they are put in fresh water. The spider crab, for instance, is so inefficient that the concentration of salts in its fluids can be changed simply by diluting the sea water in which it is sitting, and if it is put in fresh water the salts pass right out of its body. Only a few species such as salmon, sea trout and eels can move happily between salt and fresh water.

* An ion is an atom in which an electron has been stripped off or an extra one added. This means that the electrical balance between the positively charged protons and the negative charged electrons is upset and the atom as a whole has a charge, making it a positive or negative ion. It is because of their uses in ionised form that the salts in the body are known as 'electrolytes'.

Although the colonisation of the land and air offered interesting opportunities for living systems, we should remember that, even today, marine life holds by far the greatest wealth of life, phyla and species, followed by land and then fresh water. For example, the Cephalopoda, i.e. squid and the like, are not found in freshwater environments. Nor indeed are there any representatives of the Echinodermata, the sea cucumbers and urchins; it was Baldwin who pointed to facts such as these which 'suggest that only a selection of many kinds of marine animals have succeeded in establishing themselves in fresh water'.

To look for a reason for this fundamental imbalance, we need to consider the differences between sea, fresh water and air. If we assume that life originated in the sea, then there are a number of problems facing it in its attempt to spread elsewhere. The sea offers a constant environment in terms of temperature and composition; moving into fresh water and on to land involves a progressive loss of that constancy. The differences between land, water and air species raise some fascinating suggestions on the links between chemistry and life.

The story of the migration from sea to land has been dealt with more than adequately in elegant detail by a number of different authors – we refer in particular to a recent book by Colin Little entitled *Colonization of Land* (Little, 1983). The main thrust of our argument here is that some very grand designs in evolution may have operated through some very simple mechanisms: nutrition may have contributed to the movement of animals from the sea on to the land and into the air.

Some animals can live in fresh water, but they face a problem in producing their young. Modern fresh water has a very low nutrient density. Most marine animals hatch out their eggs in the form of larvae, which are intermediate forms often quite unlike the adults in appearance. They feed on the microscopic green plants and diatoms which are abundant in the sea, particularly near the surface. Fresh water contains far fewer of these microscopic plants and a lower salt and trace element content. As a result larval forms are less common among freshwater animals. For the most part, freshwater animals remain inside the egg until they have completed the larval stage and then hatch out as tiny versions of the adult.

Baldwin sums up more interesting evidence relating food to the early stage of development of freshwater species. When a bird lays an egg its shell contains not only the fertilised cell that will grow into a baby bird, but also the entire supply of nutrients it needs to reach that point and hatch out. This has to be so because the egg has no chance of absorbing nutrients from the air around it.

In the sea things are different. Not only will the young hatch out earlier, as soon as they have become larvae; they can also get some of their nourishment from the water surrounding the eggs. The new-laid egg of the cuttlefish, for example, contains only 0.8mg of minerals but at the end of its development it contains 3.3mg. Three-quarters of the minerals needed by the growing larva have been obtained directly from the sea. Sea urchins' eggs are also known to take in minerals from the water as they develop. In fresh water the density of nutrients is nothing like as low as it is in air, but is still too dilute to be of much use to an egg. This, coupled with the need to support the young through the larval stage, means that freshwater creatures must lay larger eggs than their marine relatives. Baldwin illustrates this using as his example the shrimp, of which there are two varieties, one marine and one freshwater. The marine form lays about 320 eggs a year, each with an average diameter of half a millimetre. The freshwater form lays only about 25 eggs a year but with a diameter of one and a half millimetres. This means that the volume of each of its eggs is 27 times that of the marine species.

Baldwin quotes many other examples, and although we do not have precise data on the volumes of all the eggs we can see how their number reflects the general rule that freshwater species lay larger and therefore fewer eggs than closely related marine forms. The whelk lays 12,000 eggs each year: its freshwater relatives only 20 to 100. The oyster lays 1,800,000: the freshwater mussel about 18,000. A cod fish lays about 2 million eggs: salmon eggs are quite large and considered a delicacy in Japan.

Is it possible that this enforced enlargement of eggs in freshwater species was the start of a process that later led to the development of still larger eggs by land animals such as amphibia, reptiles and birds? For them there was an added problem: each of their eggs had to contain all the water the developing embryo would need as well as all the nutrients. It had to be delivered in a sealed container that would prevent the water from evaporating but would still allow some oxygen to enter for the embryo to use. (It is worth remembering that hard shells are not a prerequisite for reproduction in fresh water. Frogs lay jelly-like spawn although the principle of the large egg or nutrient volume is still there.)

Perhaps we should consider how the co-ordinated enlargement of the egg was achieved during the exploration of freshwater habitats. Ultimately this stepwise process could well have led to the much larger, hard-shelled eggs which allowed the colonisation of the land to take place. The classical explanation would be that only those species which produced large eggs survived in fresh water.

Let us examine the differences between sea, fresh water and land more closely. There are of course great differences between birds' eggs and those of fish or even amphibia. The eggs of reptiles, which evolved earlier than birds' eggs, are different again: they do not have one yolk like birds' eggs but two sacs, one in which the embryo develops and one into which its waste products are excreted. However, the principle of packaging all the necessary nutrients remains the same. What is interesting is that, so far at least, no trace has been found of any intermediate forms. It seems to be another evolutionary leap.

It was undoubtedly the egg that allowed the first animals to reproduce out of the water, and therefore to become fully creatures of the land. Some, it is true, like the worms, remained aquatic creatures in a way, since a worm must always live in moist surroundings under the soil. In wholly dry conditions, such as exposure to the sun on a flat rock, it soon dries up and dies. Indeed it is an odd fact that many species, having given up the water, presumably for some advantage on dry land, then went back into the water; this happened with many plant species. Animals like the alligator and the crocodile return to the water to hunt their food, as did the dinosaur ichthyosaurus which made a remarkably successful attempt to imitate a fish. Penguins also pretend they are fish. Some time ago, certain land mammals occupied the land-water interface and then went back into the sea. These species which had once been land mammals became marine mammals. Why, one may ask, did animals take the trouble to colonise the land and then go back into the sea again?

In our own age dolphins and whales, mammals like us, spend their entire lives in the sea; but – and here of course is the vital difference from fish – they do not live wholly under it. They come up to breathe, and so betray the reason why land animals have one supreme advantage water animals can never have, a more plentiful and accessible supply of oxygen. Was oxygen the attraction, driving the move on to land: a move which provided us with yet another discontinuity or evolutionary jump?

A hundred litres of sea water at 20°C has about 0.53 litres of oxygen dissolved in it, and because fresh water has a low salt content it can dissolve *more* oxygen: about a third as much again. In every hundred litres of air there are 21 litres of oxygen – 40 times as much as in water. Here indeed was a driving force for the land-seeking systems, one which made it worthwhile dealing with the rigours and inconstancies of land. Remember, it was oxygen which made animal life possible and this advance was stimulated by the fact that energy can be produced eight times more efficiently with oxygen than without it. One fact that has always been apparent but seldom discussed is the converse: water contains very little oxygen. The water-living systems could not have been free to exploit the full potential of oxygen. The lowest oxygen concentrations will be in the ocean and especially at the depths where the *Titanic* lies. This means that the jump from water into air meant a huge jump in metabolic energy.

When female cod breed they lay millions of eggs in mid-ocean. The male fish swim amongst them in a frenzy of mass fertilisation. This action takes place on the surface where the hatchlings find a wealth of planktonic and other food resources. So where the cod lose out on oxygen they gain in nutrients. Although they themselves may feed in the body of the ocean, they lay their eggs where the oxygen concentration is more favourable and where the ultraviolet light stimulates the beginnings of the food chain. The surface interface between water and air, with its wind-whipped white horses, has more oxygen than is found in deep water but even richer than this is the land-water interface where the breakers smash themselves against rocks, sand and shingle.

The oxygenation of water will be at its highest, however, in the racing freshwater streams and rivers which tumble over rocks and waterfalls. The fast-flowing rivers where the salmon leap to their spawning ground contain the freshest water and are where the sturgeon lays the caviar. In the unspoilt rivers of Norway and Canada, the salmon can be seen in packs, facing upstream where the water rushes over the rocks in what must be the most oxygen-ated part of the rivers. To an observer, they almost seem to be breathing. Their eggs are not only large but are also packed with oil, no doubt as an energy store.

The next biggest jump in oxygenation is to leap right out of the water altogether. Although we think of the freshwater breeding grounds as nutrient-poor compared to the sea, their oxygen density is much greater. It might be argued that in tumbling fresh water, we see the first signs of animals beginning to take greater advantage of the high energy production from oxygen in preparation for the jump for the real riches on land.* Was it then the super-efficiency of oxygen use, first magnified by the higher oxygen content of the freshwater stream and then magnified again many times by direct access to air, that stimulated speed of growth and so the size of the egg? Surely it cannot have been as simple as that?

In fact it might well be that simple. The salmon gathers its food from the rich resources of the sea and spawns in the upper reaches of the river. It does not eat while it is spawning. It relies instead on nutrient stores built up over the previous year or more. With the increased oxygen of the fresh water, its biochemistry can now work at a more effective rate. During the time it is building the egg it can do more; it can pack more nutrients into each egg; it is driven into building a bigger egg. A rainbow trout's egg takes about the same time as a chicken's egg to hatch. Yet in less than a month the chicken has reached the weight of a two-year-old rainbow trout.

So it is just possible that the oxygen gradient was a key factor encouraging animals to move from sea water into fresh water and then on to land. The modern, highly efficient car engines will do many more miles on the same gallon of petrol than the old breed of low-efficiency machines built before the oil-crisis. If you become more energy efficient, you can go further on the same food and do more: a species adapting to a richer oxygen environment would in the same time, build larger eggs than it could have done before.

Here then is a plausible explanation for the larger eggs of freshwater species and the even larger eggs on land. This explanation can stand on its own without a selective force. That is not to say that natural selction was not at work. What is intriguing is that the jumps in oxygen availability from sea water to fresh river water to the land could explain one of the important objections raised by

* Frogs lay their spawn in still water – but they are of course air-breathing animals during their reproductive phase.

Rattray Taylor. Why are there virtually no intermediate fossils if natural selection progressed by random selection of small changes? How does one explain the suddenness with which major changes in pattern occur? In the case we discuss here, the answer could be quite simply that the leap in oxygen availability led to leaps in efficiency, which promptly dictated leaps in egg size. Any creature which developed by mutation or possessed the genetic capability to move into babbling brooks or on to land, would immediately, by force of the leaps in chemical efficiency, produce a larger egg. There would be no intermediate forms. If a species with aspirations towards the land were to find itself in an environment which released it from a pre-existing gene suppression, the move would have been rapid.

If oxygen provided the energy with which to pack the eggs, where did the shell come from? One of the rigours met by the species that colonised the land was the force of gravity. All marine organisms weigh much the same as the water they displace. The buoyancy of the sea supports them and they float without effort. They are helped by the fats in their bodies, for fats are lighter than water, mostly between 82 and 92 per cent lighter, and float on it. Seals and other marine mammals that carry a layer of fat beneath the skin gain in buoyancy; humans do the same.

The salt dissolved in sea water adds to its buoyancy: in fresh water some of that buoyancy is lost, but on land things are altogether tougher. Every land animal must support its own weight in a way that allows it to move around. The massive leg bones of the elephant bear witness to the sort of structure that weight demands and to the quantity of nutrients that must be diverted into building and maintaining it. The principal nutrient required to build bones is calcium, and that calcium had to be available on the land. By contrast, calcium availability in the marine food chain is poor. Perhaps it will not have escaped the attention of those who eat kippers for breakfast that the bones in fish are very thin. At the top of the marine food chain, the sharks use cartilaginous material where calcium-type bones might have been expected.

By contrast, the land–water interface species (bivalves such as mussels, oysters and limpets) excrete a great deal of calcium into their shells. It is a long shot, but the problem of building eggshells of thicker dimensions and rich in calcium might have simply been solved because the land-water interface provided a wealth of calcium-rich sources and hence a stepping-stone to the land. The shellfish of the seashore may well testify to the coastal and land-water interfaces as calcium-rich regions: they themselves would have offered a good source of calcium.

Filter feeders process huge quantities of sea water to obtain their food. Could the cost of this technique be the expense of dealing with surplus minerals from the water? Perhaps surplus calcium was simply dumped as excess to requirements and became shells. This dumping process occurs on wash-hand basin taps in hard water areas. You can demonstrate (at school) that these sea shells are made of calcium simply by putting one in a test-tube with hydrochloric acid. You will watch it bubble and effervesce whilst it dissolves like a vitamin C tablet. The hydrochloric acid forms calcium chloride which is soluble and the carbonate turns into bubbles of carbon dioxide identical to the bubbles in mineral water. You could dissolve the White Cliffs of Dover in the same way.

There are very few freshwater filter feeders, but reasonable numbers of snails of various types inhabit the interface. However, it is not necessary to suppose that all colonisation occurred via fresh water. The oxygenation of the sea water, constantly frothing over sand and rock, could have been a sufficient stepping-stone. After all, turtles come out of the sea to lay their eggs on the beach: perhaps they just stay that way because they prefer seafood.

The problem of gravity, at first a barrier to colonising the land, would seem to be a much more acute problem for flying animals. The colonisation of the land required a supply of heavy, solid, skeleton-building materials to support the weight previously supported by water. Birds now had to cut down their weight drastically or they could never have got off the ground; the first thing they had to do was lighten their skeleton. The colonisation of the air would, in the Darwinian scheme, be achieved by many generations of pre-birds finding that those with slightly lighter skeletons were favoured because there was survival value in the air.

In time the cumulative changes would have reached the point where a sufficiently light and appropriately shaped skeleton had developed, and the first bird would fall out of a tree and fly, or take off and so survive as the fittest. There is, however, something distinctly odd about this approach. What advantage could be gained by colonising the air? Indeed, if there was such an advantage why was it that having learnt to fly, some birds gave it up? Ostriches are not particularly good at it, nor, for that matter, are hens.

We would like to ask whether food could have been a contributing factor to the construction of skeletons of a particular type just as it was for shells? Some interesting light was thrown on this and more by (oddly enough) an encounter with some orphaned lion cubs.

Michael Crawford recalls: 'While working in Africa my family and I came across three lion cubs whose mother had either deserted them or, more likely, been killed. Knowing they would come to a nasty end, we piled them into the back of our Landrover and took them home. As we didn't have a licence to keep lions, we gave them to the Game Department at Entebbe. At the time we picked them up they were each about the size of a hairy spaniel with chubby legs and decidedly useable claws. They made curious noises of the kind you would expect to come from a none-too-pleased overgrown pussycat and displayed a mixture of tender lovingness, coyness and downright aggressive play all at the same time. We visited them from time to time and one was used in the film *Born Free*. About six months later one of them went down with hind-leg paralysis and had to be shot. A second became ill with the same problem. We left Africa at that time but assume the disease also visited the third. It was a sad story.'

It was Richard T. W. Fiennes, the pathologist to London Zoo, who pointed the way to explanation. He was interested in the hind-leg paralysis which had been described in lions at the end of the last century. While working with nutritionist Dr Patricia Scott, Dick Fiennes compared X-ray photographs of the long bones of a zoo lion with those of a wild lion skeleton in the British Museum. The zoo lion's bones were translucent and wild bones were dense. Furthermore, in the zoo lion the *foramen magnum* (the bone space through which the spinal cord descends from the skull to serve the rest of the body) was puffed up, like an arthritic joint, so that the opening was narrowed, compressing the cord. This was the explanation for the hind-leg paralysis.

Dr Scott answered the biochemical questions. Feeding cats in the laboratory on meat only resulted in a similar translucence of the bones. Meat contains a great deal of phosphorus but little calcium and both are needed for bones. Too much phosphorus prevents proper calcium utilisation. The reply from the zoo was: 'But we feed the lions hind quarters of horses which has meat and bone.' The final answer came from a knowledge of the behaviour of lions in the wild. The lion's jaw, unlike the hyena's, is not strong enough to crack the long bones of large animals, which is what they were being given. They can and do eat the rib-cage and that is where they get their calcium. They also eat the liver which is a rich source of vitamins A, D and other nutrients. It followed that our lion cubs in Uganda, fed meat only, would have been getting inadequate calcium and too much phosphorus. This would inevitably lead to a very thin cortex (the outer shell of the bone) giving them light bones. Could there be a mechanism on these lines behind the thin skeletons of mammals and birds that colonised the air?

If this argument possesses any grain of truth we should see some parallel evidence and, in fact, several parallel lines do exist. It would be logical to suppose that the hard eggshells of land reptiles and birds would have been made possible by a greater bio-availability of the elements needed to make shells on land than in water. The suggestion that there is a greater bio-availability of calcium on land is supported by the evidence in the rhinoceros and other large mammals. The rhino weighs a ton at the age of four and all the calcium needed for its massive bones is available on land from the simplest possible food resource – grass.

Further evidence comes from four sources: first from krill and fish bones. The tiny animals at the start of the marine food chain contain little calcium. Krill, shrimps and the like have chitins instead of bones. So they provide a high-phosphorus low-calcium food which may have something to do with the small bones of fish and the absence of bones in creatures like squid and octopus.

At the beginning of evolution, the marine species would have been very tiny and working towards becoming bigger. The vertebrates were fairly late in arriving and when they arrived with calcium-rich bones, they would have faced an established calcium-poor, phosphorus-rich food chain because that was the nature of soft tissue and the pre-existing species. The creatures at the top of the marine food chain, sharks, have collagenous material instead of bones.

Secondly, seagulls eat fish and lay eggs with soft shells.

Thirdly, the dolphin foetus has legs. These contract and disappear. The genetic mechanism for producing legs is present but not active. In the wild it eats fish and fish have small bones but us much as 50 per cent of its diet can be made up of squid and similar species which have no bones.

The amount of calcium in squid and octopus is very low. In the meat of salmon and cod there is 13.4 and 13.7 times more phosphorus than calcium and about the same ratio in whole squid. However, the fish also have bones, and sea water contains a significant amount of calcium (0.4g/100ml). Muscle function and movement requires more phosphorus than calcium* so at the start, animal biology was without bones and involved a high-phosphorus low-calcium ratio. Hence the low-calcium level in marine life would be a function of the biology at the start of the food chain and life. What is clear from present knowledge is that a high-phosphorus low-calcium ratio acts against the hormonal and metabolic mechanisms involved in bone growth. It is just possible that weightlessness combined with a high-phosphorus low-calcium diet may have dictated an economy in bone formation in the marine, vertebrate species. Such an economy might explain the sacrifice of the dolphin's legs. (The effects of weightlessness have been demonstrated in space exploits. The Soviet cosmonauts, who hold the world record for living in outer space, have to be carried out of their ships in special chairs after prolonged exposure to weightlessness. Yuri Romanenko who lived in space for 326 days lost about 5 per cent of his bone calcium and had probably lost up to a third of his blood volume.)

The fourth piece of evidence is provided by the El Molo, an African tribe, which relied almost entirely on fish from Lake Rudolph. Sir Vivian Fuchs first visited them in the 1930s and reported on the curious fact that they all had bent legs. Michael Crawford remembers: 'In 1965 I organised a safari to see if this was still true and to establish if the reason was a genetic abnormality. It was noted, however, that passers-by sometimes joined the idyllic life by the lake and produced children who started life with straight legs which bent when they learnt to walk. As all the children were equally affected, the reason for the weak leg bones could not be genetic.

'Lake Rudolph is a huge, jade-coloured lake and analysis showed it to have a very low calcium content. The El Molo lived by the

* Phosphate is used in the energy molecules involved in contraction. The muscle stores energy as creatine phosphate.

lake and were restricted to a diet of lake fish because of their otherwise barren surroundings; a lava desert ran right down to the lakeshore. They also caught only big fish because small fish escaped from their wide-meshed nets. The bones of these small fish would have provided the El Molo with a valuable source of calcium but by eating only the meat from the large fish, the El Molo had a diet that was low in calcium but high in phosphorus. All their weight-bearing leg bones were weak and bent: the evidence incriminated their low calcium intake. To them, their sabre-shaped legs were natural. So here is evidence of nutrition affecting shape and form in our own species. Perhaps the El Molo were on the way to becoming freshwater dolphins.'

What this story makes clear is that there is a nutritional and physical mechanism for producing light bones, and we suggest that this merits consideration as part of the evolutionary process that led to the colonisation of the air by insects and then birds. We can enjoy an interesting perspective if, for a moment, we take a lobster's-eye view. Scuttling round on the floor of the sea the lobster sees around it first the bottom layer feeders like flounders and plaice, then above them it sees the middle-level feeders like cod and sharks and finally the top-level feeders like sardines. To the lobster the fish must look as the birds look to us, and a Darwinian or Lamarckian-minded lobster might speculate on the process of competitive selection or on the 'need' that drove them to abandon the sea floor and 'fly' in the water above it. To the fish such speculations would be meaningless. It is just as natural for them to float or swim in the water as it is for the lobster to crawl on the sand. They are there because that is where their food supply is to be found, and that food combined with the DNA blueprints is capable of building the kind of creatures they are. Perhaps we should look at the flying insects, birds and the colonisation of the air in the same way.

There is another problem, which gravity poses for land-dwellers, that no fish has to worry about: falling. For a land animal whose food is found in precarious places it is a constant danger. The hedgehog is an interesting example. It has long been thought that its quills are protection against predators but studies of the animal have revealed little evidence of their being used for defence or aggression.

A British scientist has put forward the suggestion that the quills were developed so that the hedgehog could bounce. The idea of a bouncing hedgehog seems a bit bizarre, but hedgehogs apparently spend a lot of time climbing trees and walls in search of snails hiding in the ivy or in crevices, and when one of them falls – which it can hardly avoid doing occasionally – it does quite literally bounce. The stresses were calculated and it was found that no other animal of comparable size could fall from the same heights and escape injury. The quills were very effective shock-absorbers.

The evolution of the quill could have resulted from the selective advantage obtained by climbing higher and higher into the trees and rock faces in search of snails and then falling out of them. On the other hand, it might be that the hedgehog found, much to his surprise, that on his first fall, he bounced. This experience was not as unpleasant as expected and so he ventured higher and higher until he began to recognise the limit to his shock absorbers. Which of these alternatives is chosen depends really on the individual's value judgements.

However, the quills of a hedgehog and those in a bird's feathers are similar in structure. There is almost certainly a genetic and/or chemical reason why the hedgehog's quills take the shape and form that they do and are not more like feathers. If the hedgehog's quills had developed fluffy attachments it might have learnt at least to glide out of trees and perhaps, somewhat unconventionally, to fly: but it would have had to give up eating calcium-rich snails!

We have noted that calcium can be found in greater abundance in rocks and on land than in the sea or in the air. The flying insects, much loved by small birds, are not exactly a rich source of minerals. They are somewhat analogous to the crustacea (e.g. shrimps) in the marine environment in that their skeleton is composed of chitins rather than mineral-heavy bones. If, as seems likely, birds evolved with a taste for flying insects they would have evolved with light bones.

The idea of a heavy sparrow-sized animal with thick bones developing translucent bones and then, in a Lamarckian sense, 'needing' to fly is not appealing. But if light-bodied insects colonised the air first (as they did) the first birds to use them as food may have been very tiny creatures themselves. The Ley-san teal occupies a habitat in Ley-san Lagoon, Hawaii. It spends its time rushing around the marshland at the water's edge where millions and millions of flies feed off rotting debris and plants. These ducks wade through the mob, their feet making loud splatting noises on the damp ground. This action produces a dense cloud of insects in front of its face. All it then has to do is to walk into the cloud snapping its beak open and shut to capture large mouthfuls of the succulent insects.

Perhaps we forget what the world looked like before there was anything to eat the insects of the air. At the beginning of their evolution they themselves, with their tiny light bodies and genetic instructions for wing-like structures, may well have been simply blown into the air. There is no need to invent some kind of advantage. If the habitat of the Ley-san teal is anything to go by, the first tiny animals to meet the insects would have had a field day: but they would have had little calcium with which to make strong, heavy bones.

Falling out of plants or tiny trees or jumping at insects could have been a flying instructor. The humming-bird, which eats the nectar from flowers, could not have evolved until there were flowers. On the other hand it is possible that when the flowers did appear, there was already an established bird which developed a taste for their nectar. Whatever happened, it ended up drinking nectar from flowers in bushes and trees, a process which might initially have been possible from the branches of the early small plants but later required flight. Nectar, of course, contains little calcium. Similarly, the fruits of the flowering trees are potassium-rich but calcium-poor foods. The fruit-eating bats would have evolved on a low-calcium diet. Hence the eaters of flying insects, nectar, fruits and of surface habitat fish, would have had at least one factor in common: a low-calcium diet.

Funnily enough, monkeys in the zoo provided further evidence in this direction. It was traditional to feed them on fruit and vegetables but fruit and vegetables contain little calcium. Monkeys fed on this regime had serious bone problems, known as 'rubber bone disease'. Feeding them with calcium- and vitamin D-rich foods solved the problem.

There have been no investigations into just how insects as a food could have stimulated biochemistry into developing the feathers of birds or the wings of bats and there is no reason to suppose that they did. Bats are mammals and their wings contain thin, delicate arm and hand bones. The dolphin has arm and hand bones in its flippers. In both the webs between their fingers filled out the space and the bones contracted. Although the emphasis is different, bats and dolphins possess the same anatomical change. The dolphin flies in the sea and the bat swims in the air. Was this change directed by a coincident set of genetic mutations in bats and dolphins? Or was there feedback between the nutritional and physical environment resulting in a change in genetic expression and finally in an alteration of the genetic codes?

Even the evidence we have so far discussed indicates that this line of reasoning is worth pursuing. Flying insects, birds and bats would have evolved with the concept of buoyancy in air. All we are saying is that, at the very least, their food structure had to be compatible; it had to produce a light skeleton and a small, light body which could be blown into the air, then float in the winds and air currents. We know that certain foods can make the body grow faster and others make it grow slower; so far in this chapter we have seen specifically how certain foods favour and others restrict bone growth.

It is equally plausible that some species used foods which encouraged body growth. In his studies of the food habits of different species in East Africa, Hugh Lamprey coined the phrase 'biological separation'. Different species use different ecological settings from which to select their food: no two species use the same niche. For example, the giraffe eats the tops of the trees, the eland the middle layer, and the oryx the grass. You do not find another species competing with the giraffe for the tops of the trees. So at the beginning some species would have evolved with a taste for calcium-rich foods and some with a taste for different foods. The calcium-rich diet would have favoured big heavy bodies and the other light bodies.

The large-bodied birds have a poor flight ability: hens and pheasants, for example, are easy for us to catch and make excellent food. The largest predators of the air, the eagles and vultures, live on a high-phosphorus low-calcium diet and have maintained a light body to wing span ratio. By contrast, the birds which developed heavier bodies either never took up flying or were forced to give it up. Certain birds, like the domestic hen, eat much hard stony material to support a prodigious egg production. A bird that eats a lot of stones is likely to have a problem in flying!

* * *

If buoyancy separates land animals from those in the water, so too, to an even greater degree, do the enormously increased fluctuations of temperature that they have to face. The root of this problem lies in the very different thermal capacities of water and air. Thermal capacity is simply the ability of an object to hold heat, and it is measured by finding how much heat it takes to change the temperature of that object by a fixed amount. The thermal capacity of water is so much greater than that of air that it takes about 30,000 times as much heat to change the temperature of water as it does the same volume of air. That is why the temperature of the sea stays within a comparatively narrow range through winter and summer, while that of land can fluctuate between violent extremes not just with the seasons but between day and night. We have already referred to the fact that the biochemical reactions of our bodies and of the other living things of our planet can only operate within a very narrow window of the incredible galactic range.

The ultimate response to this challenge of temperature swings on land is that the warm-blooded (or homeothermic) animals control the temperature inside their bodies to remarkably close limits. Clinical thermometers have to be calibrated in tenths of a degree because that is the sort of accuracy to which our bodies work. At the time when the first animals moved from sea to land, however, this kind of control was hundreds of millions of years in the future.

The first land animals may have inhabited places where the range of temperature was narrow – and then they must have just hoped for the best. It may be that at that time the earth was in places still hot or going through a hot period, and certainly evolution on land has been associated with higher temperatures than we find in the sea today. Some fish can live and move around with a body temperature of 4°C, whereas land animals have to hibernate or go into torpor if their temperature drops to that sort of level. Fortunately for the animals the reactions that the body uses to generate energy throw off waste heat (as we discover every time we do something strenuous and sweat to keep our temperature from rising) so an animal's body temperature can rise above that of its surroundings. The larger the animal, the more pronounced this effect will be, since it will take longer for their bodies to cool through. The giant dinosaurs probably maintained quite high body temperatures, not necessarily through any regulatory mechanism but through the combination of their metabolic activities, huge thermal capacities and the sunshine.

The insects, like the reptiles, did not produce homeothermic species, though those latecomers the bees have developed an interesting communal method of keeping warm. When the breeding season starts in early spring the colony gradually generates heat, raising the temperature of the hive to around 35°C and from then on they hold it at a fairly constant level until the end of the breeding season. In spring a bee-keeper can tell how well a particular colony is doing by putting his hand on the hive cover and feeling how warm it is. On a hot day bees will stand at the hive entrance and use their wings to blow a current of air through the hive. During the winter bees go into part-hibernation: the temperature of the colony drops and only a modest amount of activity goes on at the centre of the cluster round the queen.

Even when homeothermic animals did evolve they did not all attempt to keep their temperatures constant all the time. The most extreme example of this is hibernation. In the strict sense of the word this means letting the body temperature fall to the level of the animal's surroundings, but in practice this is rare. Of the mammals, the dormouse comes closest, but most, like the hedgehog and the grizzly bear, find a hole or a cave to shelter in and maintain enough metabolic activity to hold their temperatures at a level at which the system will keep ticking over.

Apart from hibernation, a number of mammals, faced with extreme conditions, have opted not to keep their temperatures constant all the time. Those that allow it to vary are known as poikilotherms and by doing so some of them have increased their efficiency in dealing with a harsh climate. An interesting comparison has been made between the European cow and the African oryx and eland, two species of antelope, similar in size.

The comparison is an important one because of European attempts to help African agriculture by introducing European cattle into Africa. In many places this has been a disastrous failure, offering further evidence of the role of conditions. The native species clearly have survival mechanisms adapted to hot climates that European cattle lack; so much so that during the severe drought in northern Kenya in 1962 and 1963, European cattle died in thousands and those who depended on them became famine-stricken refugees; the eland, however, continued to reproduce and rear their young throughout the eighteen-month period when no surface water was available.

Dr Dick Taylor, an American physiologist, had closely studied the work of Schmidt Neilsen, one of the first to explore the mechanisms for survival of desert animals. Taylor worked in Kenya and was impressed by the way in which certain large antelopes enjoyed hot dry conditions. He captured a number of oryx and eland, built a room in which he could control the temperature and humidity, and studied the mechanisms whereby they were able to live in a hot semi-arid environment.

How did the animals deal with their water supply and survive during droughts when cattle died? The first point to emerge was that the antelopes needed to hold the same amount of water in their bodies as the cows. They had no advantage there. The next thing to look at was their kidneys. Some animals, like the desert rat, have developed super-efficient kidneys that can get rid of all the waste products passed through them by using a tiny amount of water to dissolve them and flush them out. Dr Taylor found that although the eland and oryx had quite efficient kidneys they were not in the class of the desert rat, so that could not be the answer.

What he discovered was that there was not one answer but many. Cows, like humans, maintain a constant body temperature and when it gets too hot they lose a lot of water by evaporation. The eland and oryx at some stage abandoned that idea and allow their body temperature to rise to over 43°C at midday, and at night to fall to 33°C. The loss of water through perspiration is avoided.

Next there is the water that we lose through our lungs, which can be seen when we breathe on a glass surface and watch it mist up. The eland and oryx have highly efficient lungs which extract more of the oxygen from the air than do the cow's: so, needing to pump less air in and out of their bodies, they lose less water for the gain of the same amount of oxygen.

Then there is the animals' dung. The oryx and eland, unlike cattle, produce dung as dry, hard pellets, which means that their digestive processes are extracting water from their food far more efficiently than the cow's.

The semi-arid adapted browsers like eland and giraffe also have the good sense to stay in the shade during the heat of the day and to eat at night; there is a less obvious reason for this than simply keeping the animals cool. The semi-arid browsers eat leaves instead of grass. The principle can be witnessed even in Britain where the leaves of a tomato plant will visibly shrink when the sun is burning down on them and fill out again with water in the cool of the evening. The leaves of balanites and acacia trees, on which the kudu, eland and giraffe feed, do the same. At midday their water content is around 20 to 30 per cent but at midnight it rises to 60 or 80 per cent. So by feeding at night the leaf-eating eland remains independent of surface water: it gets all the water it needs from the plants whose tap roots bring it up from the water table deep underground.

This story has a side-bearing on agricultural policy. The potential of the plants and animal species adapted to hot, dry climates and resistant to indigenous disease, has for some curious reason been consistently ignored by those who gave aid to Africa. For example, the physiological adaptations of giraffe, eland and oryx, to mention but a few, are so remarkable, and the meat of the young animals so splendid and their milks so rich, that one wonders why their use has been ignored.

Furthermore, trees, bushes, special grasses and tubers exercise their unique economy with water and nutrients with a track record worth millions of years. They have been a part of African traditional food ever since man occupied these areas. In Uganda some of the Karamajong run eland with their cattle simply because the *nyama* (meat) is so good. The Russians have been using eland in Askaniya Nova for milk for some years (Treus & Krevchenko, 1968).

When the Europeans brought their cattle into Africa they cleared the trees and bushes, shot the eland and other wildlife, and then dug bore-holes to spill the precious, underground water on to the surface, where it was exposed to the direct rays of the sun and quickly evaporated. The water tables sank and trees for some distance around were deprived of water and died. This destroyed the microclimate under their canopy which was not only cooler but was also kept moist by the gentle transpiration of water through their leaves. As a result, the sedges, herbs and edible grasses also died.

In Uganda in the 1960s the wild animals that were shot were left to rot in the sun to make way for exotic cattle. Enough meat to feed several townships for a year was given to flies and vultures. Pairs of bulldozers followed. Connected by huge chains they were marched across the once rich but now deserted plains of Ankole tearing out the deep-rooted trees. The idea was to produce grass pasture for cattle. Even here, with Lake Victoria close by and a favourable rainfall, the response of the grasses was for the succulent species to die and one could only watch as the heat-resistant and unpalatable lemon grasses took over. An interesting example of how fast a change in conditions can change the dominant species.

The disaster would have been total had it not been for the inefficiency of the operation, which paralleled its stupidity. Many of the trees regenerated from branches trampled into the ground by the bulldozer caterpillar tracks. Sadly, the wildlife, with its unique physiological adaption to hot dry climates, did not come back. One wonders why later, in the 1970s, the World Bank and the EEC, at a cost of several million dollars, set up an International Livestock Centre in Africa for more cattle development – in Ethiopia!

What the eland and oryx showed is that the evolutionary response to hot dry climates was multifactorial. It involved the mechanism for regulating temperature, the surface areas of the lungs, the absorption of water in the intestines, the permeability of the skin preventing water loss through perspiration, the efficiency of the kidneys, a metabolic response to produce a fat-rich milk, nervous system sensor modifications and an appropriate behaviour pattern even involving the shape of the mouth and tongue. Not only do all these changes work together, but the evolution of each one of them involves a number of separate steps. For all this to have evolved demands a complete set of biological rules and mechanisms, and it leaves us again with the question of whether random changes could produce such a co-ordinated programme.

Suppose, on the other hand, we were to argue that the plants of the hot dry environments encouraged the animals' ability to retain moisture by developing the integrity and function of surface areas. If that were so then the skin would lose less water; the lungs would develop their surface areas more efficiently, extracting more oxygen and losing less water; and the intestines too would lose less water. This is not a wild suggestion, because we know that plants from hot dry regions have to be able to control water loss more efficiently than plants from wet areas, so they must have a better range of chemicals and mechanisms for achieving this. Could those chemicals also be involved in the mechanism by which animals retain water?

As it happens, there is an extensive body of scientific evidence showing that a deficiency of linoleic acid, the essential fatty acid found in plants, increases the rate of water loss through an animal's skin. Conversely, the correction of a relative deficiency improves water retention. Indeed, the effect of linoleic acid upon skin water loss continues to be felt at a range of intakes which are far above what one would call deficiency. Long before we knew this question existed, the Nuffield Laboratory published several scientific papers which show that tissues from semi-arid adapted species such as eland and giraffe had significantly higher levels of linoleic acid than the same tissues from wet species like buffalo or cows.

There is a parallel here with the way giraffes feed. The long neck of the giraffe was used by Darwin as an example of natural selection. The advantage gained was obvious – it could reach food unavailable to other species. On the other hand, there is a rather interesting contradiction: female giraffe are shorter than males, which would surely put them at a disadvantage. Yet the females have the higher nutrient requirements to reproduce the young and to produce the milk.

There are also other species with long necks but which are very much smaller. The gerenuck is a semi-arid adapted species no bigger than a sheep so it is eating at a level directly accessible to cows and many other short-necked species which do not eat the same type of food. So where is the advantage of the long neck?

Perhaps the obvious is not necessarily the answer. Their long necks may not simply be a device to allow giraffe to browse beyond the reach of other animals. It could simply be a mechanism for obtaining a particular type of food. The secret of the giraffe was revealed by Robin Pellew, a Cambridge graduate working in Tanzania. He found that they feed selectively and what they eat directly supports their elongated structure. Robin Pellew found that giraffes ate only about 5 per cent of the plant species available to them. Our own studies of giraffe tissues showed that the result of this selectiveness, allied to the animal's own digestive system, provides it with high proportions of the 'essential polyunsaturated fatty acids'. Interestingly enough these include linoleic acid and also those which are employed extensively in the vascular system for the regulation of blood flow and pressure – something a very tall animal must obviously control efficiently.

Fats are also important in another evolutionary response – namely to cold. The saturated fats, the sort found in butter, have higher melting points than the unsaturated fats. Polyunsaturated margarine spreads straight from the fridge but butter does not. Being able to spread a fat at refrigerator temperatures has obvious implications for animals without a temperature-regulating mechanism and which live at temperatures as low as 1 or 2°C: namely fish. At that temperature a fish built of saturated fats would freeze solid.

Dr Bob Ackman of the Canadian Fisheries Research and Technology Laboratories has published extensive data on the composition of fats in fish caught in the cold waters off Nova Scotia and has found that they are of the most unsaturated kinds known. Dr Andrew Sinclair, working from Melbourne, has shown that fish from the warm Australian waters contain a much lower degree of unsaturation. A similar effect is found in the bodies of reindeer. The fat in their feet and lower legs, which spend much of the time in snow, is more unsaturated and more fluid at lower temperatures than the fat in their bodies.

The response to temperature is, as is to be expected, primarily seen in plants at the start of the food chain. Dutch scientists have shown that bulbs grown in a cold climate produced more unsaturated fats than the same varieties grown in a hot climate. These examples illustrate how physical mechanisms and the food chain are relevant to evolution. As animals cannot make the polyunsaturated fats themselves, it is obvious that the evolution of the entire range of North and South Atlantic marine animals as we know them could not have occurred if the food chain had produced saturated instead of polyunsaturated fats.

Gravity is another force perceived at all times by the nervous system. Lie in bed for a month and calcium will come out of your weight-bearing bones. The brain and the nervous system not only feel, they also control through stimulation, regulation of blood flow and an armoury of hormones. It is common knowledge how fat influences body shape and buoyancy. It is quite feasible that shapes could therefore respond to differences in buoyancy as well as to differences in nutrients. Again, specific nutrients could relate to physical responses including the control of water loss through neural and skin structural mechanisms.

Nutrition is an extremely young subject and there is much that we have yet to learn. But at least the evidence we have discussed offers a mechanism whereby chemistry could determine or stimulate the direction of evolution and even explain the discontinuities of evolutionary grooves and leaps in egg size. We claim no more than that these mechanisms deserve attention.

In this chapter we have looked at some of the ways in which life systems of various kinds evolved in sympathy with the conditions in which they found themselves. First the simple prokaryotes occupied the planet for 1.6 billion years. Here if ever was the greatest discontinuity in evolution. No matter how much random variation or struggling for survival occurred, nothing could happen until the blue-green algae had excreted enough oxygen to saturate available receptors and raise its level to the point at which oxygen-using systems – animals – became possible. It was the presence of oxygen which stimulated and made possible the new wave of evolution. Since its inception, only 500 or 600 million years have passed during which time evolution has worked its way in discontinuous jumps from the trilobites to the graptolites and the fish. Then, from the many species that evolved in the sea, a few began to colonise the land. The plants colonised the land first and so they moved into rich, virgin territory. Their evolution exploded in the midst of plenty. The amphibia were followed by the reptiles, the dinosaurs, the modern mammals and finally man.

The multicellular systems were woven around the nervous and vascular systems, driven by oxygen and super-efficient production of high energy from food, with the energy dissipated in movement. Paradoxically, this new phase of evolution was using oxygen-sensitive building materials. Despite all the protective mechanisms available, no solution was found. This is not to say that natural selection did not play its part. But chemistry and environment were also crucial: they both stimulated and determined the directions.

The same truth continued to hold as the animals developed into many new species and colonised the land and the air. Was it the physical and chemical stimuli of oxygen and minerals which made faster-growing and bigger eggs with hard shells, reversed the process to make soft seagull eggs and switch off the dolphin's information on hind legs? Was it a taste for insects made of chitins that enabled the birds to fly and the air to be colonised? Whatever happened, we suggest that the animals initially colonised the land, not in a struggle for survival but in the midst of plenty. Maybe the air was colonised in the same way.

Much later that marvellous, dangerous species man appeared. In the chapters that follow we shall be looking at the evolution of our own species; could man have been driven forward by the demands and influences of chemistry and physics? We shall also ask how this idea might apply to human history and our present situation.

We conclude here with some remarks by Darwin that show how he himself would have argued against those who deny the influences of conditions or food, and who see no need to find any evolutionary mechanism other than random mutation and the survival of the fittest. The quotations come from *The Origin of Species*.

> Several writers have misapprehended or objected to the term natural selection. Some have even imagined that natural selection induces variability, whereas it implies only the preservation of such variations as arise and are beneficial to the being under its conditions of life.
>
> In looking at many small points of difference between species, which, as far as our ignorance permits us to judge, seem quite unimportant, we must not forget that climate, food etc., have no doubt produced some direct effect.

> [On the origin of domestic races of animals and plants] . . . Changed conditions of life are of the highest importance in causing variability, both by acting directly on the organisation, and indirectly by affecting the reproductive system.

> Our ignorance of the laws of variation is profound. Not in one case out of a hundred can we pretend to assign any reason why this or that part has varied . . . Changed conditions generally induce mere fluctuating variability, but sometimes they cause direct and definite effects; and these may become strongly marked in the course of time.

If the greatest events in evolution history were actually determined by physics and chemistry, then it is indeed most likely that the smaller events will have been similarly fashioned. How some of those 'direct and definite effects' came about, and how they became not only 'strongly marked' but irrevocably built into the species concerned, will be a major theme of the next chapter, in which we investigate the great chain of dependence by which all advanced life-forms are linked.

6 The favoured predators

Watch television and your eyes will process 25 images a second. If you think that is remarkable, consider the cat at night. As you step into the dark and crack your shin on the garden stool, the cat's eye has turned night into day and it is darting through the undergrowth in search of unsuspecting nourishment: in the meantime your own eyes are struggling to adapt to the dark.

At night the iris in the cat's eye opens wide to let every scrap of light impinge on its highly sensitive retina. Shine a torch at it and, at a speed almost faster than you can see, the iris clamps down to a narrow slit. It might be supposed that this rapid response is an immense advantage to the cat. Perhaps cats without this capability perished. On the other hand, it is highly unlikely that there were people, at night, shining electric torches at cats as they were evolving! Although to our human mind such an ability would seem to be an advantage, we must remember that darkness comes on gradually: sunlight is not switched on and off. Hence it is difficult to see what advantage such a fast response mechanism conferred.

The Nikon-like automatic-exposure setting of the cat's eye could have been created by selection but could also have been simply the development of a highly integrated mechanism in response to *a priori* encouragement by appropriate chemistry. Whatever the stimulus was, it was not just the chance putting together of a highly flexible iris with night vision and an appropriate centre in the brain. Had it been a random putting together of all these separate things, we would expect to find highly flexible irises in near-blind species and we don't.

The eye is one of the first protrusions to be thrown out by the developing brain. Once the ground-plan of the eye is established, it is not difficult to understand how consistent exposure to appropriate chemistry could encourage that protrusion to develop whereas deficits would restrict it. Going along with the higher development would be the accessories like the articulated iris, the control of which simultaneously demands similar chemical inputs. It does not look like a random business.

The basic guts of the eye design with its retina and iris is common to all higher mammals but the large carnivores contrast most dramatically with the corresponding herbivores. This rigid separation of the species is not readily explained by chance and such discrepancies were seen by Rattray Taylor as a cardinal failure of selection theory. It was one of the points which made him conclude that there had to be another directive force in evolution.

Even human eyes are amazingly efficient in dealing with different degrees of light. If you look across a valley on a clear night in the country you can easily see a 60-watt light through an uncurtained window six miles away. Put your eye just two feet away from the same light and, though you would find it uncomfortable, you would be able to read the writing on the glass. It is a measure of the range of stimuli to which our eyes can respond that the strength of light from the distant source is more than a billion times weaker than from the close one. The cat and the owl can do even better.

It is easy to imagine the advantage good night vision gives to a predator, and classical evolution theory has always supposed that this explains it easily enough. What still needs an explanation is why the animals who are preyed on by the cat family have not also developed good night vision? Why should it not be just as advantageous for an antelope to see a leopard in the evening shadows, and so avoid it, as it is for the leopard to see the antelope? And if evolution proceeds through random mutations why has this development not occurred in a single herbivorous species?

There is a clue in an unexpected place. After the First World War Denmark exported all its butter to earn foreign exchange and for home consumption made do with margarine. There was a high incidence of blindness among Danish children and it was this that led to margarine being fortified with vitamin A. At that time little was known about vitamin A. This vitamin, which is also called retinol, is now known to be essential for the growth of nerves in the eye and for night vision. It is made in some (but not all) animals by the conversion of beta-carotene in plants (carrots and dark, green leafy vegetables like spinach) and in some coloured fruits to active retinol or by eating other animals which have already done the work.

The first sign of a vitamin A deficiency is failing night vision, and one of the factors that could possibly have encouraged cats to develop such good night vision (we shall be looking at others later in this chapter) is that they are able to get a particularly good supply of it. But here we encounter a curious paradox. Most animals can make their own vitamin A from beta-carotene in plants. Cats, which have such a strong need for the vitamin, cannot do that. They have to obtain it preformed from their prey, who have it stored in their livers. What possible advantage can it be to the cats to be unable to produce a vitamin they need so badly?

Oddly enough there is an advantage; although strictly speaking it would be more proper to say an opportunity is created. Every function that an animal has to carry out demands an appropriate mechanism and organisation within the animal. Now there is a limit to the level of organisation that any particular animal can achieve and also to what it can pass on through its genes. In a way this is like the limited disk space in a computer, and if any function can be delegated to another organism it leaves the disk space free to perform some new function or to perform an old one better. The more complex the organism, the more DNA it needs to carry all the information required for its construction and function. Mammals have 1400 times as much DNA in each of their cells as a simple bacterium, and despite the immense amount of information that can be encoded in it there is still a limit to the use that can be made of it. For some reason, the total amount of DNA is the same for all cells in the same animal and, even more interestingly, from one mammal to another. There seems to be a fixed amount that is feasible as a mammalian design.

So with the cats: if they had to synthesise their own vitamin A in large quantities from beta-carotene it would take up a significant amount of their disk space. They may therefore be better off delegating the process to their prey: at one meal they can take in the whole supply of vitamin A that their victim has accumulated over a lifetime, and do so at low cost to their own organisation.

Delegation of this kind is what seems to have allowed the whole hierarchy of animal life to be built up. Every species has certain compounds which it needs but cannot make for itself out of simpler compounds. We now call these compounds vitamins or essential amino and fatty acids. We cannot, for instance, synthesise all the amino-acids that go to make up our protein; those that we cannot make – the 'essential amino-acids' – we must get ready made from the organisms that we eat, as we must do with vitamins. The secret code for making these compounds is in the DNA of plants and bacteria. The more complex life systems evolved later than the simpler systems and that basically is what evolution was about. The vegetable and bacterial systems had to be there, building up not only carbohydrates, amino-acids and vitamin C but also highly complex molecules like vitamin B12, which animals later came to use. The new animal forms of life could be relieved of that function and could deploy their powers of organisation in other ways. It is plausible that an availability of these interesting molecules in food made new developments in biology possible, not unlike the way in which oxygen made animal life possible.

This growing dependence on ready made building units is the guts of the evolution of simpler systems to higher organisations. It is also nutrition. It is quite plausible that, at the beginning of animal evolution, it was this mechanism which created the phyla. Such a mechanism would at least explain the appearance of the phyla, mostly at the outset and with little change thereafter.

Large dimensions of computer programming for vitamins and essential amino-acids could be rewritten to produce new designs and the upper limit for information storage could be reached without the need to code for the long-winded synthesis of these complex molecules. Once the vacated disk space contained the new meaningful codes, the possibility for any further dramatic change, such as we see in the different phyla, would be minimised. Simultaneously, the seeds for what was to become 'nutritional science' were sown: they took 500 million years to germinate!

The science of nutrition basically concerns the elaborate system by which advanced forms of life delegate the production of some of their biochemicals to less advanced forms. It is useful to look at some of the things it has discovered. In 1887 the Dutch government established a research laboratory in Java to study beriberi, a very serious disease common throughout the east which brings on muscular weakness, nervous disorders and finally death. In the wake of the revolutionary discovery of bacteria by Pasteur it is not surprising that the disease was originally thought to be due to infections. However, a young army doctor named Christian Eijkman was placed in temporary charge of the research. When a new hospital manager decided that feeding the chickens on hospital scraps was pilfering, Eijkman had to buy rice from the market for them. He then noticed that they developed a similar condition to beriberi. It was soon found that the chickens responded to different diets and that feeding them polished rice produced the disease, while feeding either unpolished rice or the polishings effected a rapid cure. Later in 1905 Professor Pekelharing, who had originally been in charge of the Java laboratory, followed Eijkman's lead, gave up the search for a bacterial origin and published the results of conclusive experiments demonstrating that beriberi was related to a factor in the food: vitamin B1.

At that time it was believed that food consisted only of protein, fats and carbohydrates, and of these protein was considered to be of the first importance (the name is derived from the Greek *protos*, first). It had been discovered that carnivores could live off a diet of protein without fat or carbohydrate, and there was a growing appreciation of the need for protein for growth. But in 1906, the year after Pekelharing published his results, an English biochemist named F. Gowland Hopkins (later Sir Frederick) also came to the conclusion that there was more to food than protein, fats and carbohydrates.

> Scurvy and rickets are conditions so severe that they force themselves on our attention; but many other nutritive errors affect the health of individuals to a degree most important to themselves and some of them depend on unsuspected dietetic factors. I can do no more than hint at these matters, but I can assert that later developments of the science of dietetics will deal with factors highly complex and at present unknown (Hopkins, 1906).

In 1912 Hopkins rose to fame by publishing the results of experiments showing that when rats failed to grow on a carefully purified diet, he could add to the diet a factor isolated from food, which restored growth. This led Hopkins immediately to the idea that specific constituents other than protein were needed for life. This was the beginning of an exciting period in the biological sciences, when chemistry gave birth to biochemistry and when the basic knowledge was provided for the present understanding of nutrition.

In the years since then not only beriberi but scurvy, rickets, pernicious anaemia and vitamin A blindness have been largely eradicated except among some very poor peoples. It is easy to forget how recent much of our present knowledge is, but within living memory those diseases were widespread. An elderly American surgeon told how he became convinced of the vital role of nutrition. As an eager medical student on the ward rounds he recognised the symptoms of pernicious anaemia in the patient lying in bed in front of the class and, thinking to impress his teachers, he blurted this out. Far from being impressed, his tutor, with a dark expression on his face, took a firm grip on his arm and guided him through the swing-doors outside. There he spoke in stern, deliberate terms: one did not mention pernicious anaemia in front of a patient because the name was as good as a death warrant. Only two years later, in 1926, G. R. Minot and W. P. Murphy, following a lead from experiments on dogs, discovered that feeding raw liver effected a dramatic cure. The dread of pernicious anaemia vanished almost overnight although it was not until 1948 that the active principle in the liver, vitamin B12, was isolated and crystallised. An excellent account of this history of nutrition is to be found in a book by Professor J. C. Drummond and Ann Wilbrahim called *The Englishman's Food* (Drummond & Wilbrahim, 1969).

Man is not the only animal that must obtain B12 from its food. Indeed it seems that the need is common to all mammals, and what were once thought to be exceptions have turned out to be further examples of higher animals delegating the task of production to lower organisms. The richest source of vitamin B12 is liver. It used to be believed, for example, that vegetarian animals did not need vitamin B12, but then it was observed that rabbits kept in a cage with a wide wire mesh for a floor, developed symptoms of B12 deficiency. The reason, it turned out, was that rabbits normally eat their own faeces. In the wild they leave their soft droppings at the mouth of the burrow and when they return after foraging they swallow them. While the rabbits are away bacteria have been busy producing B12 in the droppings.

Grazing animals also use bacteria to produce B vitamins for them. Cows keep a bacterial flora in their stomachs, where they produce B12. The fermentation that makes yoghurt and a number of other foods generates a significant amount of B vitamins. It is worth considering that B12 is a large molecule which takes a large number of steps to synthesise. A whole series of enzyme systems is required and so a great deal of DNA coding has to be devoted to producing them. There would have been a big saving of disk space in using bacterial DNA to do the job. On the other hand, animals need not have evolved from something which could make B12. Maybe the availability of B12 made certain other chemical reactions feasible.

Because the need for B12 is common to all advanced animals, and because their successors have all inherited it, we can safely assume that mammals evolved with this need, or that B12 played an important role in stimulating advanced animal design. It is unlikely that the different species would have developed the need independently, just as it is unlikely that many different families of animals would have independently developed a need for the same essential amino-acids to build their proteins. The amino-acids essential for one species are also essential for all other mammalian species. It is all too much of a coincidence.

So in investigating the requirements for certain nutrients that we and other species have, the science of nutrition is also exploring the process of evolution.

The march of progress for individual aspects of nutrition has, however, been dependent on the advance in technology for measuring them. The need for protein for body growth was recognised over 100 years ago and the early part of this century was filled with a cascade of information on vitamins.

The most important new group of chemicals investigated in recent years is that of the *essential fatty acids*. These are used to build structural lipid in a similar way in which essential amino-acids build protein. Structural lipids construct cell membranes.

The point about these lipids is that they are quantitatively the most important structural component in the brain and nervous system and the second most important in all other soft tissues. They account for 50–60 per cent of the materials used to construct the brain. They also have special significance in other 'membrane-rich' systems, the most relevant to our discussion being the network of blood vessels.

The word 'essential' is here used in the same sense as in essential amino-acids: man and other species need them to build tissues and to reproduce but cannot make them. They are literally essential: indeed they may well be of more importance to our health and performance than even protein. The bad press that fats in general have had is confusing and must not be allowed to blind us to that fact. The confusion can be dispelled by explaining that there are two different types of fat in the body: (i) structural; and (ii) storage. The problem with our food today is that there is far too much storage fat in it. It is this problem that has attracted the bad press. Unfortunately little information is disseminated to people about the importance of the essential structural lipids.

An understanding of the different types of fat provides an important key as to how the big cats developed better nervous systems and night vision compared with the cows. We also believe it is relevant to the question of why *Homo sapiens* has a big brain.

Histology is the study of living tissue and a histologist pursues his science by looking down a microscope at thin slices of tissue which have been stained by chemicals to reveal particular components of the cell. Histology was developed for two reasons, first to describe the anatomy of the cell, and secondly to identify the pathological processes in it so as to help in the diagnosis of disease. One of the first signals of a pathological process identified by the early histologists at the end of the last century was the infiltration of fat into tissues. It was this observation that led them to recognise *two* completely different types of fat.

The first of these fats was 'visible' to the naked eye and is the sort that hangs in bulges round the waistline, its function being to store energy. It was this sort that histologists found infiltrating diseased tissues. The second was 'invisible' and was clearly of a different kind because it could only be seen with special staining techniques. This second sort was found *inside* the cell and is the structural lipid. Cell membranes are not just envelopes but active parts of the cell, arranged like an orderly production line: making things, breaking down others and transferring raw materials – the nutrients – from one part to another as needed. It is on the external and internal cell structures that the integrity and organisation of cells depend.

Although the existence of these two kinds of fat was recognised at the turn of the century, the relevance of their fatty acid composition was unknown until 1930 when Dr G. O. Burr and his wife, who were working in Minnesota, discovered that they were needed by rats. They were investigating the role of viatmin E in ovulation, the process by which the female egg descends from the ovary and is made ready for fertilisation.

The Burrs noticed an annoying variability in the reproductive power of the rats they were studying and, as vitamin E is a fat-soluble vitamin, they put this down to the fat in the rats' diets carrying varying amounts of it. To check this they raised a colony of rats on a purified, fat-free diet; the result was a deficiency syndrome totally different from that caused by a lack of vitamin E. Searching for an explanation, the Burrs fed the rats with different fats and oils and discovered that the substance that prevented the deficiency syndrome was a fatty acid known as linoleic acid. They therefore called this an 'essential fatty acid'. It was not until 28 years ago that a group of American research workers reported that human babies given milk formula without the essential fatty acid which the Burrs had described, developed very nasty skin lesions: these were repaired by feeding the essential fatty acids.

To understand the nature of the fatty acids we need to look inside their molecules. The heart of the molecule is a set of carbon atoms bonded together to make a daisy-chain. In a saturated fatty acid, each carbon atom in the chain is attached to its neighbours by one of its four bonds, with all spare bonds along the chain occupied by hydrogen.

In an unsaturated fatty acid two adjacent carbon atoms are joined to each other by two bonds. The more carbon atoms there are in the molecule that are double-linked to each other in this way, the more unsaturated the fatty acid. A *polyunsaturated* fatty acid is one in which there are two or more such double links.

Fats which are built largely with saturated fatty acids – saturated fats – are hard at room temperatures. Candles used to be made from the saturated fat found in fattened beef. They form rigid structures. By contrast, fats made with polyunsaturated fatty acids are liquid. Cells use a different balance of the rigid and fluid lipids depending on cell type and function.

The higher the degree of unsaturation in a fatty acid, the greater will be the fluidity of the lipid. From the hard tallow candle to the stiff pat of butter, to soft margarine, to liquid vegetable oil, the progression is simply the addition of double bonds. A single extra bond makes the difference between a solid candle, stiff butter and liquid corn oil.

Although animals can tolerate quite wide differences in diets, there are limits and too much saturated fat will disturb the balance and lead to what is known as hardening of the arteries: the relevance this chemistry has to human health will be discussed later.

The important point about this difference in the physical nature of the fatty acids is that it offers biology a wide range of constructional materials with quite different properties to meet different structural requirements. For example, the insulating outer sheath of a nerve is made of myelin, which uses saturated fats to confer rigidity and make sure messages do not short-circuit and go to the wrong place. By contrast, the flexible lining of the artery has a high requirement for polyunsaturated fatty acids.

To provide an even greater variety of building properties, there are fatty acids with different degrees of polyunsaturation. The highest degrees are found in regions of fast action such as in signal transmitters and receptors in the brain and in the nervous and visual systems. The reception points, the synaptic junctions, handle large numbers of incoming and outgoing signals and therefore have a high demand for flexible structures.

Our own perception of where we are, which way our toes are pointing, what our fingertips are touching, whether we are hot or cold, the control of our muscles and breathing, requires transmission, reception and interrogation of signals. The brain draws its conclusions by cross-referencing a bank of stored information. The sight of a large lobster in a rocky pool at the beach could, through this process, lead to curiosity or fear and withdrawal in someone who had never seen a lobster; in the more experienced, to a complex pursuit and lobster thermidor for supper.

All of this nervous system and brain activity depends on a range of neurotransmissions acting at millions of sites where the structures use highly unsaturated fatty acids. The best example of the specialised use of polyunsaturated fatty acids in highly active sites is in the photoreceptor, where the most polyunsaturated fatty acid (docosahexaenoic acid) accounts for 60 per cent of the fatty acid used.

This difference between the relatively saturated nature of the nerve fibre's insulating material and the highly unsaturated nature of the signalling structures is an elegant example of the way in which the different properties of individual fatty acids can be used by nature for specialised purposes. Neither protein nor carbohydrate structures have this same wide physical versatility. It is an important point to which we shall return several times that the brain is the organ with the heaviest investment in this type of biotechnology and the vascular systems come a close second.

All animals including ourselves have to obtain these polyunsaturated fatty acids from food. The two essential fatty acids (EFAs) in plants occur in an interesting complementary pattern in foods. Linoleic acid is found in seeds and the other EFA, known as alpha-linolenic acid, in their leaves and other green parts. (This difference of origin, as we shall see, has an important bearing on evolution.) The body cannot use one in place of the other. Linoleic acid is the parent member of a family of fatty acids derived from it, and now referred to in the health food industry as the Omega-6 family. Alpha-linolenic is similarly the parent of the Omega-3 family.

The Burrs discovered the role of linoleic acid while investigating fertility and it is not only the structural role that is important in reproduction. Mammals in general make from it a group of hormone-like substances called prostaglandins and leukotrienes which among other things regulate blood flow, the immune system and the mammalian reproductive process. Prostaglandins are so powerful that special ones are used in minute doses to induce delivery of babies when difficulty is being experienced. Certain of the leukotrienes (LTC4, LTD4 and LTE4) are roughly a thousand times more powerful than histamine. The work done describing these compounds led to Professors Sune Bergstrom, Bengt Samuelsson and Sir John Vane being awarded the Nobel Prize in 1982.

A deficiency of EFAs has also been associated with atrophy of the testicles. In addition it has been found that linoleic acid is crucial for skin and cell membrane integrity. We noted previously how animals that live in arid places and therefore need to have skins that are less permeable to moisture have higher levels of it in their body tissues than those living things in wet areas.

Originally there was some doubt as to whether both EFAs were needed or just the seed type (linoleic acid). Whilst studies on rats were conclusive for linoleic acid they were not so clear for alpha-linolenic acid. Evidence on chickens now points to a need for both types, especially for higher functions. In addition there is now a large body of evidence, recently reviewed in a book by Dr W. Lands, which indicates that the balance between the two types is a regulator of blood flow and the tendency of the blood to clot. Consequently the optimum balance of fatty acids in the diet is now seen as crucial for the prevention of heart attacks.

Dr Ralph Holman of the Hormel Institute, Minnesota, USA, and one of the world's leading experts in fatty acids, was asked to study an infant who developed neurological disturbances while being fed by a drip. It turned out that the drip contained only one of the EFAs, linoleic, the seed acid. When alpha-linolenic acid was included in the drip the neurological signs disappeared. More recent evidence from Dr K. S. Bjerve and his colleagues in Norway has supported this finding in a number of patients (Bjerve, Mostad & Thoresen, 1987).

Most especially, the EFAs are essential for the development of the brain. Originally people thought that the brain could make all its own fats but in the mid-1960s two researchers, Klenk in Germany and Jim Mead in the USA, reported not only that the brain contained essential fatty acids, but that they were of the highly unsaturated type.

It was the oil companies which provided the technological breakthrough. When they suspected people were diluting high quality oil with cheaper stuff, they developed techniques for separating the individual components of oils to 'fingerprint' oil wells. The technique (gas-liquid chromatography) was immediately adapted to biomedical research. Thanks to the oil companies we now have an intimate knowledge of fatty acids in the body.

The fatty acids used in the *brain* are *uniquely different* to other tissues. The 'parent' plant EFAs with 18 carbon atoms with two and three double bonds are converted in the animal at a slow pace to corresponding fatty acids with 20 or 22 carbon chain lengths and four, five or six double bonds. It is in this way that docosahexaenoic acid is contructed for the retina. Both parent plant fatty acids and their long chain derivatives are used for animal structural lipids. This means that animal tissues are made with more liquid fatty acids, which is what an animal needs, since the parts of its body, unlike a plant which is rooted to ground, are in constant motion – legs, hands, heart, arteries, blood, kidneys, nerves, digestive tract, everything except bone. At its simplest, animals have a higher requirement for liquidity because they move about whilst plants stand still.

The distinguishing feature of the fatty acids chosen for the brain is that whilst other tissues use a mixture of chain lengths, *only* these long chain, more polyunsaturated fatty acids are used to make the brain cell lipids. Nowhere is the extra flexibility more important than in the brain. The human brain has a thousand billion cells, each one of which makes 6,000 or more connections with other cells.

The process whereby the parent plant fatty acid, linoleic acid, is converted to arachidonic acid (20-carbon fatty acid with 4 double bonds), and alpha-linolenic is converted to docosahexaenoic acid (22-carbon fatty acid with 6 double bonds) is shown in Figs 8 and 9. These long chain, highly unsaturated fatty acids can be called the *neural* fatty acids.

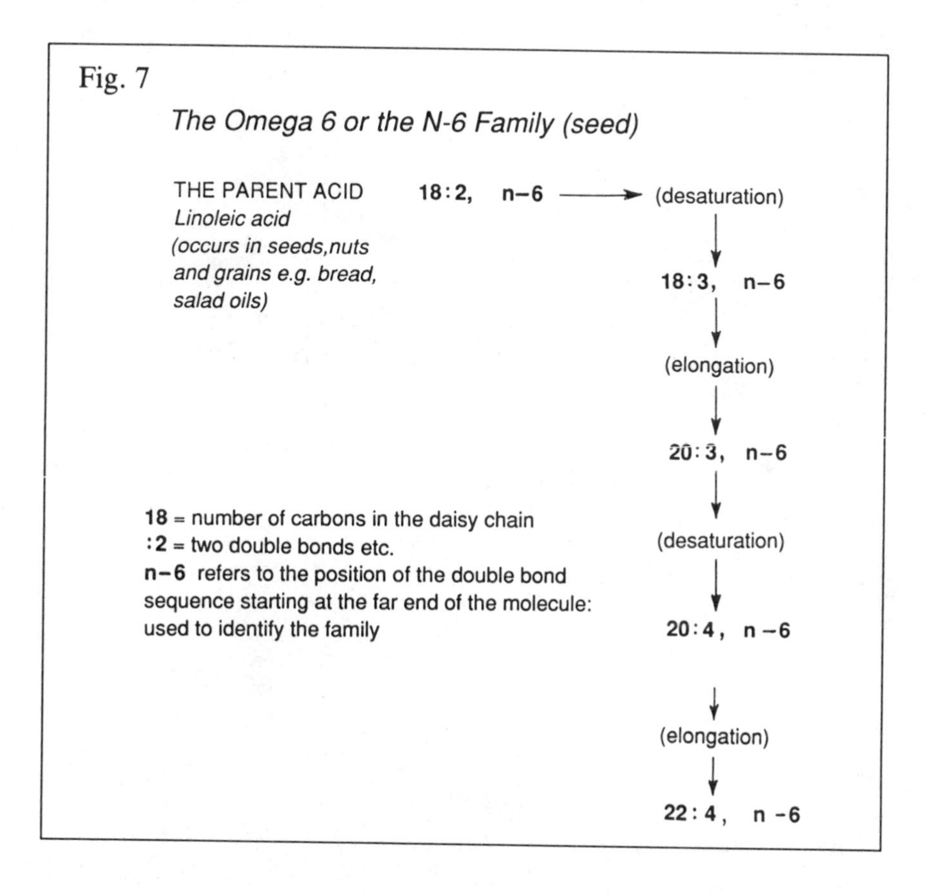

At the Institute of Zoology we asked a very simple question. 'Did different species which eat different foods use different EFAs to build their tissues?' A study of 42 species of mammals from zebras and leopards in Africa to tiny viscachas from South America and dolphins showed that although different fatty acids were used for muscles and livers, all species used the identical profile of fatty acids in their brains. This held true despite wide differences in the fatty acids in food (dolphins or cows) or in other parts of their bodies.

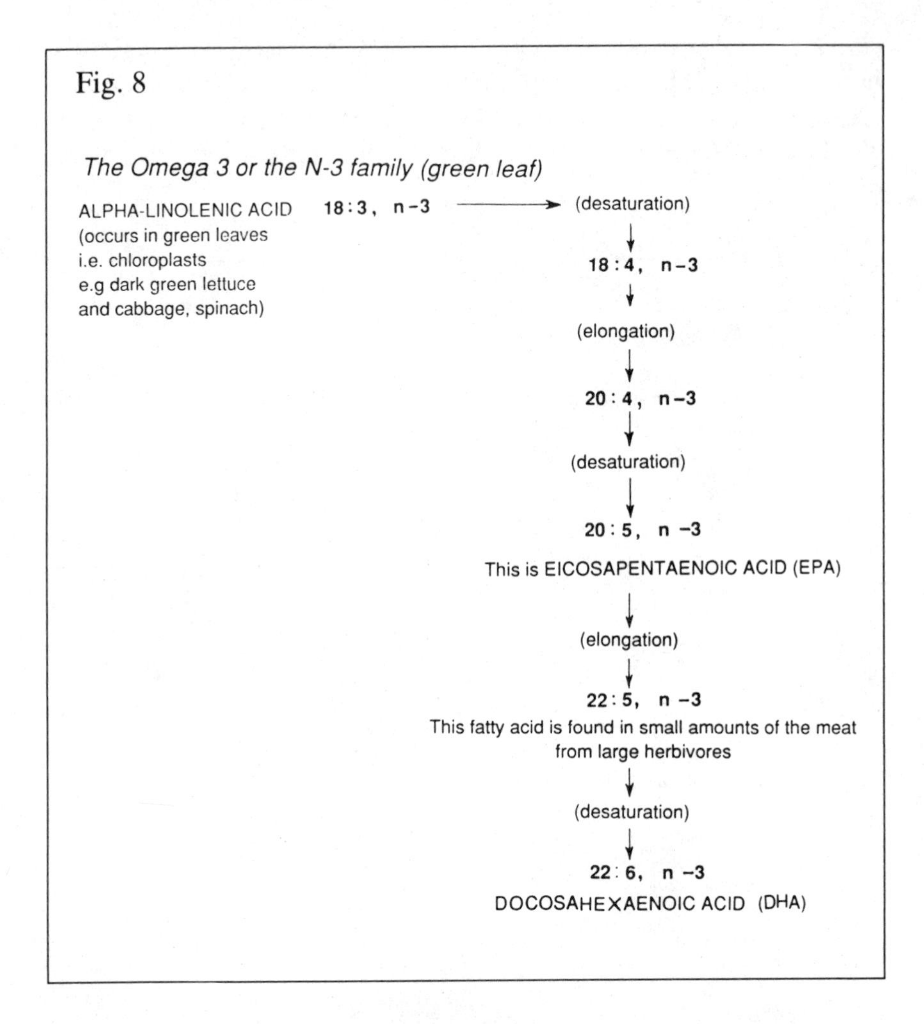

The variable in the brain was not its fatty acids but its size. Indeed, it was as though the availability of the special long chain fatty acids used in the brain was somehow related to the extent to which the brain was developed.

Were all species able to make the neural fatty acids from the starting point in plants? In all small mammals like mice, rats and tree shrews, there was evidence of ample conversion. However, in the large herbivores like cows, buffaloes and giraffes, which eat leaves and seeds, we could find plenty of the original linoleic and alpha-linolenic acids in their tissues, but the conversion process to the neural fatty acids was not completed; it seemed to peter out.

Table 1 Percentages of fatty acids in the livers of different species

	18:2,n–6	20:3,n–6	20:4,n–6	18:3,n–3	20:5,n–3	22:5,n–3	22:6,n-3
Small mammals							
tree shrew	9.4	1.6	18.0	0.2	1.5	1.69	16.0
viscacha	20.0	0.7	11.0	0.6	1.6	1.1	22.0
large herbivores							
zebra	47.0	0.3	4.2	1.6	0.5	1.0	0.2
giraffe	12.0	0.3	12.0	2.1	2.1	9.3	0.8
wart hog	22.0	0.6	12.0	3.5	3.7	5.0	1.6
elephant	10.0	4.5	10.0	2.8	3.8	11.0	0.6
small carnivore							
cat	5.3	1.9	16.2	0.9	0.9	0.6	20.0
hyaena	6.8	1.2	13.8	1.2	0.2	2.8	12.0
large carnivores							
lion	2.1	0.8	14.0	0.9	1.2	5.9	5.8
dolphin	1.7	0.2	13.0	0.1	8.2	1.3	11.0
omnivore							
human	7.2	2.1	19.0	0.3	1.5	4.2	9.6
marine muscle							
salmon	2.5	0.0	4.2	2.5	8.2	1.8	47.0
cod	0.8	0.1	1.6	0.3	11.0	1.9	41.0
crab	1.3	0.4	5.5	0.2	33.0	0.7	19.0

Note: Desaturation of the parent EFAs occurs mainly in the liver. The above data was obtained from single analyses using the most recent data we have available on liver ethanolamine phosphoglycerides. The choline phosphoglycerides displayed the same loss of 22:6,n–3 in the large herbivorous land mammals. Only the carnivorous species showed evidence of an ability to collect significant amounts of long-chain derivatives of the EFA. Data on the ethanolamine phosphoglycerides of salmon (freshwater caught), crab and cod muscle are presented for additional comparison in the context of food.

The production of the most polyunsaturated fatty acid, docosahexaenoic acid (22:6,n–3) was most affected. The tissues of the large herbivores accumulated only small amounts of these neural acids, whereas the big cats which ate them had many more.

This was interesting because carnivores have more sophisticated nervous systems and presumably a higher demand for neural fatty acids. Were they perhaps better at making them? Surprisingly, we found the opposite. John Rivers and Andrew Sinclair reported another peculiarity of cats in 1975: cats did not convert plant EFAs into neural fatty acids (Rivers, Sinclair & Crawford, 1975). Others have found that this same principle operated in carnivorous fish. This inability is similar to the inability of the cat to make vitamin A from the carotenoids of plants. It relies on the efforts of its prey!

Once again, the economy of this method is clear. The process of building up these long-chain fatty acids is tedious. Any animal that can bypass even a part of it can switch off the mechanism, save on effort and time and possibly on genetic 'disk space'. The same argument applies as with vitamin A: the direct supply of preformed neural fatty acids could allow development of biological systems dependent on or stimulated by these nutrients.

Here is a mechanism to explain both the higher development of the nervous system in the carnivores compared with the herbivores, and the rigidity of the evolutionary lines.

The ready availability of preformed neural acids and vitamin A in the flesh the carnivores eat offers nutrients needed specifically to support more sophisticated brains, nervous and optical systems. They can afford the luxuries of widely adjustable irises in their eyes, along with the nerves and central control systems they demand. The herbivores cannot. Further, the retinas themselves are directly affected. The retina appears very early during embryonic life as an extension of the developing brain. The sensors in it, the nerve endings that can receive and respond to the photons of light entering the eye, are, like the brain itself, built of proteins and structural lipids. We have already seen that a very high proportion of the structural lipid in the photoreceptor is docosahexaenoic acid. Dr W. E. Connors of Oregon, has found that monkey infants deprived of these Omega-3 fatty acids lose visual acuity, which provides experimental evidence in favour of this mechanism. It now seems that both docosahexaenoic acid and vitamin A are essential components of night vision.

This, then, poses a question for the role of advantage as a determinant of evolution. The advantage of better vision and brains would have been just as great on the side of the prey as on the predators. These lines are held not because of selection based on advantage but because of chemistry. The predators could simply get more of the essential nutrients because the herbivores were building up supplies for them. Even if a cow's genes mutated to produce the data for a carnivore's visual system, it could not build it: it does not have enough raw materials.

Let us now put the story another way. As with the increasing availability of oxygen leading to more efficient use of food resources, are we here seeing evidence of a greater availability of the building materials leading to a more advanced building? Are we seeing evidence of nutrients or substances in the form of vitamin A and docosahexaenoic acid stimulating the development of the visual and nervous systems in the cats and owls but restricting it in cows and rhinos? Is this evidence of substrate-driven evolution?

We suggest that what gives the carnivores better eyes also gives them better brains, and it is consistently true over the whole range of species that carnivores are cleverer than their prey. The spider has a more complex pattern of behaviour than the fly. The old male lion, standing upwind of a herd of gazelle and roaring and urinating at the same time, is showing a far more complex pattern of behaviour than the gazelle who break off grazing to run petrified in the opposite direction, straight into the claws of the lionesses waiting behind the bushes.

Think, too, of the vast difference between the buzzard and the pheasant. The cock pheasant looks very pretty but all he can do is fly in such a straight line that he has evolved into a very successful target for a two-legged animal with a double-barrelled shotgun. The buzzard can control its flight so precisely that it can stand still in the air by sensing the forces of wind and gravity and fluttering its wings in response. Its visual acuity is such that it can see a mouse moving in the grass 50 or 100 feet below, and its judgement so rapid and precise that it can fold its wings, drop like a stone, calculate the exact moment to open its wings while falling at nearly 60 miles an hour, set the angle of the wings to the air to break the speed, extend its legs, open its claws, snatch the mouse and leap back upwards into the air without even touching the ground.

An owl has a night vision sensitivity which is ten to 100 times better than ours, and our night vision is not that bad. Moreover, it can use its stereoscopic hearing to pin-point a mouse under the snow without even seeing it. Owls have been known to dive headlong from a tree into a bank of soft snow, and emerge almost instantly with a mouse clutched in its mouth. Their ears can detect the subtle high frequency sounds made by a mouse eating; sounds too high-pitched for any human ear but enough for the owl's sophisticated sensor and computer system to guide it blind through the air and snow with deadly accuracy to the mouse's warm, safe hiding place beneath the snow.

The 'computer programming' needed to control such achievements is extraordinary. The one act of the buzzard's daily routine is more complex than the hardware and software needed to send an intercontinental ballistic missile from Nebraska to Siberia.

The paleontological evidence tells us that throughout evolution the brains of the carnivores have been consistently in advance of the corresponding herbivores. But if the reader doubts that carnivores are more intelligent than herbivores then the argument can simply be based on the anatomical differences. Comparing the anatomy of the leopard and the kob immediately brings the contrast into stark relief. Whilst the kob's hands and feet are in the shape of hoofs, the leopard has articulated claws. The business of using a soft foot for sensing and for controlling the claws, demands a greater investment on the part of the brain and the nervous system than does the use of a hoof.

Random mutation and natural selection offer no clear and convincing explanation for a process that in terms of nutrition appears inevitable. Random selection does not explain why the carnivores and herbivores split into two streams; why these species, living in the same habitat, divide into two functionally and morphologically distinct groups; why herbivores have poor night vision and consistently less developed nervous systems. Most especially, random selection does not explain why herbivores and carnivores are stuck in their grooves: nutrition or substrate-driven change does.

This nutritional evidence answers Rattray Taylor's eighth problem of the departure from randomness created by the rigidity of the evolutionary lines. They can be explained simply as dependence on developments from the food chain. The carnivores and herbivores are held in the grooves by nutritional differences.

We can take this line of argument a step further by looking at the way all mammals first come into the world, carnivores and herbivores alike, and so gain a greater insight into our own species. The brain is the earliest organ to develop. Seventy per cent of the total maximum number of brain cells that anyone ever has were built inside the mother during foetal life. Inside the womb the foetus 'eats' the food that the mother supplies; this food has previously been processed by the mother's own digestive system, including her liver. Just as a carnivore eats the end-products of another animal's efforts, so a foetus is nourished entirely by the end-products of its mother's work. There is more to it: between mother and offspring there is a placenta. We studied the concentration of nutrients on the maternal and foetal sides of the human placenta and discovered that the placenta was selecting the long chain neural fatty acids used in the brain from the mother's blood, and passing them on at higher concentration to the foetus. We can see a foetus – even of a vegetarian animal – as a kind of super-carnivore. The process of biomagnification applies to other nutrients as well, but in the case of the EFAs it specifically selects the neural, and not the parent plant EFAs.

From this evidence one would predict that, as most of the brain's development takes place in the womb, long gestations should favour brain development, and so should a slow rate of foetal growth, since the nutrients will have lower demands made on them by the rest of the body. This turns out to be so.

The placental mechanism for concentrating the nutrients used specifically in the developing brain offers an explanation as to why egg-laying species have tiny brains. Instead of a steady flow of blood from the placenta continuing over the weeks or months of pregnancy, the growing creature inside the egg is presented with a once-for-all package. Then, once it is clear of the egg, it must take its food from the surrounding environment without having it first concentrated inside its mother's body, concentrated a second time by the placenta and then again through its mother's milk.

We would maintain that it is this inferior system of nourishing their developing progeny that condemns the egg-laying species to perpetual inferiority of brain power. No fish, bird, snake or other reptile has ever evolved a large brain. However big a bird's egg (and they are, of course, enormous compared to those of fish) none has ever cracked open to reveal a big-brained chick. The ostrich can weigh as much as a fully grown human: its brain is the size of a child's fist.

Marsupials occupy a place between the egg-laying species and the placental mammals. The kangaroo has a very short gestation period of a matter of two weeks or so before the joey moves into the pouch. They end up with bigger brains than egg laying species, but smaller than placental mammals. When mammals reached South America they ousted the marsupials, leaving Australia as their haven.

The gestational period of the cow is the same length as that of the human. A new-born calf, however, weighs as much as a teenager. At two years old a calf can weigh 200kg but have only 350g of brain: a human infant at the same age weighs about 15kg and a brain weighing about 1.00 to 1.2kg. Its brain is three times the size of the calf's, and in proportion to total body size it is around 60 times as big.

The comparative data would be weak without some experimental evidence demonstrating nutritional effects on brain size and function. Experiment does more: it shows just how critical the period of brain growth is. Professor John Dobbing and Dr Elsie Widdowson reallocated newborn rat pups so that some mothers had 14 and others only 4 to feed with their own milk. The result was that the pups in the larger litters became microcephalic (they had small brains), whereas those in the small litters had much larger brains (Dobbing, 1972). No matter what they did afterwards, the difference remained fixed. The rat's brain develops mainly during the period the pups are suckled. Dobbing concluded that there is a 'vulnerable' or 'critical' period of brain development during its early growth. Any deficits or distortions that occur during this period cannot be corrected once it is over, and this shows in humans who are born blind or educationally subnormal due to some distortion in development: the defect cannot be put right later. The developmental period is critical. It is during this period that brain growth can either be stimulated or stunted and the effect is permanent in the lifetime of the individual.

Several laboratories have now shown in experimental animals that specific alterations of dietary EFA intakes can influence brain development, synaptic function and, as we have seen, visual acuity. In 1973 Dr Andrew Sinclair, working at the Institute of Zoology, tried feeding female rats on a diet with a low concentration of EFAs. Without any they failed to reproduce at all so he gave them enough to allow reproduction but not enough for safety. The experiment was continued over three generations and resulted in low brain cell counts in the third generation pups.

In Milan Dr Claudio Galli similarly showed that if lactating rats were fed on a diet deficient in EFAs their pups sucked deficient milk and suffered permanent learning defects which could not be reversed by giving them a correct diet after weaning (Galli & Socini, 1983). Like Dobbing's experiments, the EFA undernutrition during brain development led to permanent deficits.

Fats have another special characteristic: they are stored in the body. They are passed from mother to infant, and specific effects of changes in dietary fatty acids are delayed by the buffering capacity of the store, but later can be multi-generational in their effect.

Dr Galli's work also underlines the importance of another source of nutrition for the developing young: milk. Although no milk contains the high concentrations of EFAs found in blood leaving the placenta (especially neural fatty acids), it still does contain some and keeps the 'carnivorous' feed style going after birth. (As a proportion of the calories in human milk, the EFAs are present in a larger amount than the essential amino-acids.) Work at the Institute of Zoology has shown that species like nesting rodents, which leave a large part of their brain-cell division to be completed after birth, have higher proportions of the neural acids in their milks than those born ready to go.

There is a similar contrast between cow's milk and human. The young cow does a lot of growing but will end up with a small brain. Its mother's milk has three or four times the amount of protein found in human milk and five to seven times the mineral content – protein for fast body growth and minerals for bones. Human milk, on the other hand, has less protein but, by contrast, has six to ten times the amount of EFAs essential for postnatal development of the nervous and vascular systems. Gorilla milk has more linoleic and alpha-linolenic acids and fewer long chain neural acids than human milk. In practice, the milk of each different species is uniquely tailored to its postnatal requirements. Biology and food have long worked together to shape the different designs.

Here also, is the answer to the other conundrum about the herbivores. The first was their failure to develop good night vision: the second concerns hoofs. If hoofs were developed to give them speed of flight, how is it that the fastest four-legged animal is a soft-footed carnivore? The cheetah can hold a speed of 100kph for 60 seconds. Of course one can see the advantage of paws and claws to a lion: you cannot easily bring down a running buffalo by dragging at it with a set of hoofs. In that form the argument is convincing. Natural selection would have provided these species with claws. It is when you turn it around that it fails. If hoofs give speed why cannot the gazelle, zebra, giraffe or buffalo run as fast as the cheetah despite their hoofs and despite the fact that the extra speed would have given them an inestimable advantage for survival in

Darwinian terms? It would be hard to convince a cheetah, racing across the ground, that it would do better if it let you fasten a set of artificial hoofs over its paws!

The understanding which we are now developing in biochemistry and nutrition offers a different reason for hoofs: not speed but economy. Hands or articulated claws need a vascular system to nourish them and to supply chemical energy to the muscles that move them. They need nerves and synaptic junctions to send and receive messages and they need corresponding parts in the brain to deal with those messages. Hands and claws need, in short, a big addition to the two systems that depend on long-chain, essential fatty acids: the carnivores can get them but herbivores cannot.

The horse provides a good example because in its earliest form, about 50 million years ago, it did have toes, four on its front feet and three on its back feet, but as it evolved these were reduced first to two and then to one. In his delightfully titled book *Hens' Teeth and Horses' Toes* Stephen Gould expresses his dissatisfaction with the traditional explanation of evolution as 'the summation of small changes that adapt populations ever more finely to their local environments' (Gould, 1983). Yet here, in the evolution of the horse's hoof, is surely a convincing demonstration of a theory which, at least in its negative form, can hardly be denied. In that form our argument is simply this: whatever the mechanism that brings about evolutionary change, no new life form can come into existence if it needs a type or quantity of food that it cannot get. If there is anything strange about that very obvious thought it is only that its significance appears to have been missed by so many evolutionists, past and present.

Some understanding of what happened to the horse can be gleaned from a disease that afflicts human beings. Peripheral vascular disease, sometimes associated with heart disease or diabetes, can cause the capillary circulation to degenerate. An important sign is a 'clubbing' of the fingers, which appear to shrink and thicken. Tiny blood vessels at the ends of the fingers, which have a large surface area relative to their diameter, can no longer be properly maintained. They collapse, thrombose, or in some other way are put out of action. The tissues they should support are affected, cells that have reached the end of their lifespans cannot be replaced, and the diseased part becomes shorter and thicker as inactive material accumulates in it.

It is not difficult to see the relevance of this to the horse. We have discussed before how the embryonic development of the brain, the heart and blood vessels takes place at the earliest moment. Similarly, the evolution of the brain and the vascular system go together and, biochemically, they use similar fats. The fast growth and food pattern of the horse do not favour an extensive vascular or nervous system; this would have become ever more true as the evolving horse grew larger. By comparison with the other large herbivores it is biochemically the worst off.

At its first recognisable appearance the horse was no bigger than a dog, and in that form it had toes. It can be suggested that the larger it grew, the more demands were made on its food supply and the less there was to spare to build vascular and nervous systems in proportion to its new bulk. Hands and feet demand a huge investment in nervous 'hardware'. In the human species, whose hands are developed to the point where they can perform the most delicate micro-surgery or execute a Liszt rhapsody on the piano, the volume in the brain occupied by the nerve fibres and cells serving the hand is out of all proportion to its apparent size. The same is true of the vascular network. Since the horse had to economise somewhere, the toes and fingers were a sensible sacrifice, bringing about a disproportionately large saving in return for a seemingly unimportant loss.

The importance of this saving is clear if we look at the horse's brain. It takes up a mere 0.05 per cent of the animal's total weight (compared with the 2 per cent taken up by the human brain), and from the fossil record it appears that, using this relative size as the measure, the horse's brain has shrunk as its body grew larger. As N. W. Pirie has pointed out, small mammals like squirrels have roughly the same relative brain size as man, and it is significant too that none of them have developed hoofs: all have toes or fingers.

The horse is as pure a vegetarian as they come, whereas the cow unexpectedly turns out to be partly carnivorous. The technique adopted by the straight herbivores, like the horse and the zebra, is simply to process vast quantities of vegetable foods. They are selective feeders, using green leaves to provide the protein and minerals they need and the vegetable fibre to push it all through at speed to make excellent manure. By contrast, the ruminants such as cows, sheep and goats, hold the vegetation they eat in their stomach where a beautiful symbiotic relationship exists between resident bacteria and protozoa and the beast itself. These micro-organisms can digest the cellulose of plants – an impossible task for other species as they do not possess the digestive enzyme, cellulase, which the micro-organisms do. Hence the bacteria break down plant fibres; protozoa eat the bacteria; these are filtered in an additional stomach and then passed on to a 'true stomach' and digestive system. Here, the microbes and protozoa provide the cow with a high protein diet: as protozoa are tiny animals, the cow and other ruminants are in this sense carnivores! Why then do they not have sophisticated nervous systems?

The answer lies in the fact that the micro-organisms – the bacteria and protozoa – are reproducing without access to oxygen. There is little or no free oxygen in the cow's stomach, so how do they burn the carbohydrates to make energy for growth? Instead of combining hydrogen and oxygen, which is the way aerobic systems produce energy, they rely on the energy derived only from breaking down quantities of cellulose and carbohydrates in the plants. The large amounts of hydrogen produced in the process are then 'dumped' in any willing molecule.

At this point, the understanding of the nature of polyunsaturated fats becomes rewarding. The essential fatty acids are polyunsaturated, which means they have double bonds from which the hydrogen is missing. These unsaturated bonds make excellent hydrogen acceptors. In this way the micro-organisms get rid of their hydrogen, produce masses of proteins but in so doing, destroy much of the EFA turning them into saturated fats. Seeds, especially from bushes and trees, may escape this process. Being oil-rich, they tend to float through. Protected seeds and incomplete hydrogenation means that some EFA molecules escape. Those that do are precious and used for cell membranes; there is certainly not enough to spare for the body fats which is why beef fat is full of non-essential, saturated fats. The carnivorous cows gain the advantage of a rich supply of protein but the cost is the irrevocable and simultaneous destruction of the EFAs needed for brains and hands: hence a huge body and little brain.

It may be that horses originally had (and perhaps still have) the genetic information to make a large brain but lack the building materials. Protein and mineral deposition goes hand in hand with body growth. However, the capacity to produce long-chain neural lipids does not keep pace and their concentrations in tissues falls away as animals get bigger: consequently the brain loses out. The small mammals cope but in bigger animals with fast body growth, brain growth simply does not keep pace. Hence there is a universal contraction of the size of the brain relative to the body it serves, as well as a loss of large functional zones.

Of the mammals, the large herbivores are the worst affected. However, the secondates (such as cows and lions) are worse off then the primates (monkeys). The primates have much larger brains relative to their body size than the secondates. They have gone down a different pathway, that of slow growth, long gestation and lactation; and they do not destroy their EFAs in their stomachs. The biochemistry shows that the primates as a whole are much more richly endowed with the long chain EFA derivatives than the secondates, and their milks have much less protein and fewer minerals. So the group differences in relative brain size of the egg laying birds and reptiles, the marsupials and then the mammals, and the differences between the herbivores and carnivores, whether they be birds or mammals, can be explained by the flow of nutrients in the food chain. We shall be looking at the significance of this fact to *Homo sapiens* later.

What we have been looking at in this chapter is a contrast in strategies that animals of different sorts adopt to get nourishment. The strategy could result in substrates driving change in different directions and explain, in simple terms, the puzzle of how rigid 'lines' were developed and maintained. One strategy led perhaps to a more specialised development whereas another led to regression. In the long run there is a lot to be said for not becoming committed to an overspecialised lifestyle. The fact that *Homo sapiens* uses what is perhaps the most diverse food selection pattern, and that, in his own eyes, he is by far the most successful of the primates, may be a good advertisement for compromise and diversity.

7 Dinosaurs, elephants and whales

There may well be an interesting similarity between the present crises of elephant survival and the problems of the dinosaurs of the past. Sixty-five million years ago they, and many species associated with them, went out of existence so rapidly that theories of global catastrophe have been put forward to explain the suddenness of their extinction at the end of the Cretaceous period. One of the more controversial made headlines when Luis Alvarez and his colleagues at the University of California suggested that the earth had collided with a huge asteroid. This would have produced a vast cloud of dust, which blasted off from the point of impact, drifted through the atmosphere right round the world, blotting out the sun and cutting off its life-giving energy. The effect would have been similar to the 'nuclear winter' that is supposed to freeze the world after all-out nuclear war. The impact of the asteroid could also have tilted the earth's axis, which again would have meant dramatic changes of climate in different parts.

The evidence for this collision is the existence of an anomalously rich and narrow band of elements of the platinum group, such as iridium and osmium, in the rock strata at a depth that corresponds to the time at which the dinosaurs became extinct. The surface of the earth contains little iridium but interstellar material is rich in it. The narrowness of the band of rocks containing it suggests an instantaneous event.

However, Alvarez's fellow-countrymen Charles Officer and Charles Drake of Dartmouth College, New Hampshire, have put forward an alternative hypothesis that the iridium band is volcanic in origin, brought up from the interior of the earth, which is much richer in iridium than the surface. Alvarez did not consider a volcano to be a possibility because it would have meant an eruption some thousand times greater than the biggest known to us, the Krakatoa eruption of 1883. However, as we have seen, violent volcanic activity was characteristic of the earlier days of the planet's history. More recent studies suggest that the iridium band is not as continuous or as sharp as would have been expected; it has been deposited over a period of 100,000 years. Whilst volcanic activity diminished as the planet cooled, there is evidence of late and intense volvanic activity in the Deccan Traps of India, one of the largest piles of lava on earth. They cover an area of more than half a million square kilometres and the volume of lava has been estimated at about a million cubic kilometres. The activity occurred between 60 and 65 million years ago, which is about the right time for the final extinction of the dinosaurs and the formation of the iridium belt. As Peter Smith said in the *Guardian* (18 April 1986), 'Until the crucial piece of evidence is found (if it ever is), the debate will continue . . . In the meantime, none of this explains exactly how an asteroid impact or volcanic activity could produce mass extinctions.' Nor indeed has a crater big enough to explain such an impact ever been found.

The problem that catastrophe theories set out to solve is why 75 per cent of other species were wiped out along with the dinosaurs. Certainly this does sound calamitous, but in fact it does not need an explanation in terms of asteroids smashing into the earth or volcanoes erupting. We need only think about what is happening today to understand with painful clarity how easy it is for one species to displace and eliminate many others.

In the last few centuries 5000 species have become extinct. *The Red List of Threatened Animals*, published by the Union for the Conservation of Nature, contains a long, sad list of species either recently extinct or endangered. The tiger and the elephant hang on precariously. The greatest of all, the whales, have been offered a dubious last minute reprieve from being eaten into extinction. Less well known, the aye-aye in Madagascar, the sole surviving representative of a whole primate family, has been reduced to the point where it is feared that fewer than 50 still survive. The gorilla has been driven back into a few forest habitats. The human primate population has exploded to 5 billion in an instant of biological or geological time. This population explosion has brought the world to a state where only a few hundred of those magnificent primate cousins of ours can still find room to live. Species after species have been extinguished, some hunted for food, more of them simply forced out of the world by the sheer pressure on space exerted by the expanding population of the dominant animal, *Homo sapiens*.

As recently as Roman times, hippos and lions seem to have been living where London now stands. Their remains were found, with a happy touch of the appropriate, when a corner of Trafalgar Square was being excavated in 1962 to build the Ugandan High Commission.

In fossil records of the future, the 2000 years from then to now will be an undetectable instant, yet in that brief time the great experiment of mammalian evolution has been telescoped towards extinction by man. We, of all peoples in history, should find it easy to see how the dinosaurs could have devoured some of their fellow animals, eliminated the habitats of others and driven so many species to extinction before they brought the final ruin down themselves.

It is possible that volcanic action on a large scale produced a climatic change and the layer of iridium in the soil, and might not this, or a burst of cosmic radiation, have been mutagenic to plants? There may also be a simpler explanation, which is that the dinosaurs may themselves have been responsible. Apart from their voracious appetities, the weight of the giant herbivorous reptiles must have been destructive. First, they would have needed considerable quantities of food – for a 40-ton dinosaur could have eaten 46,000,000 kg of plant life each year. Then the trampling of their huge feet would have destroyed much of what they did not eat and inhibited its regrowth. As their numbers grew the time would come when even the giant sized plants of that time could no longer keep pace with the demands. This period was one of the hottest and it may be that the dinosaurs created their own hothouse effect. The Malthusian end of Phase 4 had come.

The leaching of the minerals from the soil would have added another dimension not generally considered. Even today something like 39,000,000 tons of minerals and salts are washed annually by rain into the oceans. We see the effects even on our historic buildings. One litre of rain water saturated with carbon dioxide will dissolve one gram of lime from cement and stone.

We must remember that at the beginning the planet's surface was rich in chemicals providing a fertile soil for plant growth. The fact that plant life was characterised by giant ginkgos and ferns testifies to this fertility. It was an age of giants and the animals that lived on the giant plants became, in time, giants themselves. However, the young and fertile planet was already showing signs of ageing when the dinosaurs reached their peak. The Nile carried minerals and elements, washed by rains from Ethiopia and Central Africa. This transfer of richness to the Nile Delta created the Granary of Rome and, until recently, nourished the Mediterranean sardine grounds. (In recent years the Aswan Dam, by retaining a high proportion of the silt, has shut off much of this process of fertilising the Mediterranean upon which the sardine fisheries depended.) This same process was operating during the reign of the dinosaurs. The dinosaurs themselves would have contributed by eating the plants and defecating and urinating: in this form the minerals and trace elements would have been more easily washed away. A dinosaur's faecal output would have been quite prodigious. It would only need the population growth to reach its exponential phase combined with a continuous undermining of the soil fertility for the setting of the time bomb to run down to zero.

Once that Phase 4 critical point was passed, the acceleration of the destructive process would have been rapid. It is equally interesting that there are still ginkgo survivors: indeed, the Viale Baptista in Modena, Italy, and the road which circuits the emperor of Japan's estate in Tokyo is also lined with them. However, they are wizened affairs compared to what they once were: Bonsai ginkgos!

During the reign of the dinosaurs the planet had been clothed in ferns and allied species: today there are 15,000 of them and 250,000 flowering plant species. Some giant ferns can still be found but they are usually in new habitats created recently from volcanic activity. Indeed, many of the volcanic hillsides in New Zealand are clothed in ferns which, if not giant in dinosaur's eyes, are certainly giant to us. The final piece of evidence is that with the 'giantness' having been washed from the land into the sea, it may be no coincidence that the only vaguely comparable giants living today are in the sea: the blue whales.

Just how rapidly a large animal can destroy its own habitat when the amount of space available to it falls below the critical point can be seen today in the plight of the East African elephant. In the days of the Protectorate of Uganda, National Wildlife Parks were established to encourage tourism and conservation. One of the localities chosen was around Murchison Falls where the whole weight of the Nile squeezes itself violently through a 19-foot cleft in the rocks. (Legend has it that a man who jumped the gap could possess any woman of his choice.)

The region had been cattle country until, at the turn of the century, a rinderpest epidemic imported by cattle swept through Africa and destroyed most of the livestock. The area round Murchison Falls was left as a mixture of forest, scrub and open grassland, and this, together with the presence of the Nile and the Falls, recommended it as a site of scenic interest with open country for the animals to be readily seen. The new park was home to buffalo, hippo, crocodiles, hartebeest, kob, waterbuck, giraffe, bush-pig, wart-hog, lion, hyena, leopard and of course elephant. Many of the animals moved in and out of the area, except for humans who were moved out. However, a source of meat is a valuable commodity and settlements sprang up round the edge of the park, where a combination of growing populations and influxes of refugees from various disputes in other regions resulted in a cordon of people, very happy to make use of the chance of any meat, leather or ivory that might pass nearby. The elephants used to migrate from Mount Elgon in the east of Uganda to the Nile in the west, but running this gauntlet became more and more hazardous and they soon gave up and settled down, trapped in the park.

By any standards the park was a good size, with 2000 square miles on the south bank of the Nile, but once the elephants were penned inside its boundaries the presssure on its woodlands and forest began in earnest. At first there was a swift increase in the number of elephants but soon it became clear that the woodlands could not support such populations on their own. The elephants were quickly turning the forests into grassland, eating some of the trees, pushing some over for exercise and stripping the bark off others which die as a result in a process called ring barking; the loss of even a small amount of bark can spell death for a tree. Eventually it was noticed that the woodland of the park was giving way to grassland and although there was plenty of grass, some of it six feet high, the elephants' health was deteriorating.

Studies on the elephants showed that there was a lowered fertility, longer calving interval, increased neonatal mortality, reduced growth rate and delayed sexual maturity. Frequent mouth infections produced large bony abscesses on the jaws with pus draining to the outside. Dr Sylvia Sykes, who was working at the Nuffield Laboratory of the Institute of Zoology at the time, studied the pathology of these elephants. She reported that the elephants living on the grasslands suffered to an extraordinary extent from hardening of the main artery, the aorta, while those on the scrubland suffered far less and those in the forest not at all. The elephants, initially confined to the park by the actions of men, were now destroying their own food supply and bringing about a fatal change in their environment.

The story was being repeated in Kenya as well as Uganda and so the last remains of the largest of our land mammals was showing just how a species could become extinct – and how quickly. In a paper presented to the Zoological Society of London in 1966, Dr Richard Laws (who was later to become Director of the British Antarctic survey and Secretary of the Zoological Society of London) and Ian Parker wrote: 'The evidence presented suggests that the extinction of many elephant populations over the next fifty years is a strong possibility in the absence of population and habitat management.'

The solution was cropping the elephants but there was powerful resistance from conservationists and from those who felt the parks would lose their key attraction. However, the health of the elephant populations declined so rapidly that even the tourists began to recognise their sickness and a cropping programme began. In Kenya it was too late to save hundreds of elephants from dying in front of the colour supplement cameras from sickness and from starvation.

The natural habitat for the elephant is woodland. The possibility that the elephants might adapt to meet the new grassland conditions was never suggested by Laws and Parker. The evidence of cataclysmic extinction because of habitat destruction was too painfully obvious. Similarly, no amount of adaptation or struggle for survival could have seen the dinosaurs through this same crisis.

Confirmation that there was a difference between the nutrients provided by trees and grass was supplied by a group of American veterinary researchers who found that the tissues of woodland elephants were richer in vascular essential fatty acids than were the tissues of grassland elephants. A parallel study from the Nuffield Institute reported the same findings when comparing woodland with grassland buffalo: it is a similar story to the difference between the giraffe and grazing species.

It is not hard to see how the dinosaurs, some of them far larger than elephants, could have borne at least as heavily on their environment. Once their numbers reached the critical point, the third and fourth phases of the five-phase evolutionary cycle would have followed their inexorable course, with pressure of numbers leading to irreversible environmental change, the destruction of the ecology, food supply and final extinction.

But then, with an equal inevitability, the fifth phase would have followed the collapse of the dinosaurs. The new conditions, mortally unsuitable to the old forms of life, were appropriate to new forms of a quite different kind, and in due course those new forms grew up to enjoy them. Of all these newcomers, the most important belonged to that revolutionary new family to which we ourselves belong, the mammals; and of all the things that distinguished the mammals from any other animals of the same or earlier times, the most important was their brains.

What triggered off this remarkable new development? Was the inherent potential for the mammals already there simply waiting to be unleashed? At the least one can be sure that conditions had been changed and were now appropriate to quite different life-systems: they would have favoured a new range of plants. Such a range did appear at just about this time. The giant vegetation of the previous epoch was replaced by the evolution of a new type of plant: the flowering plants which produced protected seeds.

It would be a far-fetched coincidence if, at the same time, there arrived on the scene animals who depended specifically on a new nutrient found in rich supply in their seeds. Yet this is what happened.

The nutrient is linoleic acid, essential for the reproduction of the mammals but not, interestingly, for the first vertebrates, the fish. Marine phytoplankton and algae produce alpha-linolenic acid and/or its higher Omega-3 derivatives. This family of essential fatty acids is synthesised in the chloroplasts hence its appearance in the phytoplankton, algae and green leaves. The first animals to evolve did so in the sea and emerged into a food chain in which these Omega-3 fatty acids dominated the nutritional scene for lipids. Fish, and probably the first vertebrates, depend on this Omega-3 family for reproduction and growth.

The mammals, however, emerged at a time when the huge green leafy plants were at the end of their reign and the gentler, flowering and seed bearing plants were emerging in their wake. These new plants introduced protected seeds and hence a concentrated source of linoleic acid into the land food-chain. The mammals which then evolved are, as far as science can tell us today, dependent on linoleic acid and its Omega-6 family for their reproduction.

Furthermore, the long chain derivative of linoleic acid (arachidonic acid) is one of the principal fatty acids used in the brain. Previously it had not been particularly abundant. But the brain is consistently built with the same fatty acids and importantly, the balance of Omega-6 to Omega-3 in all species so far studied is 1 to 1. The difference between species is not its composition but the extent to which it is developed. Could the availability of the Omega-6 fatty acids in a concentrated form have led ultimately to a mechanism whereby a larger brain could have evolved?

We have seen how the placental system is a key to foetal growth in mammals. It may again be no coincidence that the placenta is a highly developed blood system. Linoleic acid, its long chain derivatives and prostaglandins are crucially important for the health of the blood system and proper blood flow. The connection with the physiological actions of linoleic acid and these many facets of mammalian biology are too much to be a coincidence. Was the advent of the seed bearing plants and their nutrients ultimately the change which made possible the evolution of the mammals and then the big, human brain?

This new mammalian dependence on linoleic acid is a striking feature. It seems that just as oxygen coincided with the evolution of animals, and the evolution of higher animals became dependent on the provision of essential amino-acids and vitamins, we now have, with the arrival of seeds, the introduction of a new chemical building block coinciding with the emergence of life systems making essential use of it.

Once we begin to ask what the connection was between nutrient and user, whether the presence of that nutrient could in some way have influenced the arrival of the mammals, we get into very difficult ground. The standard neo-Darwinian answer would be that the series of mutations that produced the mammals was purely random. Similar mutations might well have occurred many times in earlier ages but since the resulting animals would have been unfit to survive in an environment that did not contain their essential needs they would have immediately died out. There is nothing to quarrel with in that account, as far as it goes. But does it go far enough? Is there a more positive influence at work? An active driving and directive force? Here we can approach the problem indirectly: there has certainly never been a laboratory experiment in creating mammals, but there have been experiments that could be significant. Some of them have been performed on bacteria.

Most bacteria can make vitamins for themselves, and also most of the essential amino-acids that go to form protein. If some of those bacteria are exposed to radioactivity they can be made to mutate. That much is expected, but what is unexpected is that if the bacteria are irradiated while being kept in a medium that contains one of the amino-acids or vitamins they normally make for themselves, some of the mutants that emerge will have lost the ability to synthesise it. Furthermore, those mutants may be at a positive advantage, and a fascinating experiment has shown the process in action. Zsmenhof and Eichhorn looked at a mutant of the bacterium *P. subtilis* which could not synthesise the amino-acid histidine. What they found was that if the mutant was placed in a medium that was rich in histidine along with the normal bacteria of the same type, then the 'deficient' mutant would grow faster. The reason is thought to be that the mutant can economise; it does not need those parts of its own nucleic acid or protein that would normally control the synthesis of histidine, and therefore it can save on the biochemicals needed to make them. To use our earlier analogy again, it is economising on disk space and on building materials, and so the presence of histidine encourages its replication.

Another interesting experiment has recently been carried out on viruses, in which it was found by two different New York research workers, Nancy Evans and Gabriella Santoro, that certain prostaglandins could affect the 'expression' of viral genes or their transcription. Similarly, Peter Reed, of Aberdeen, Scotland, has shown that prostaglandins switch on muscle protein synthesis. This is exciting data, and its relevance to evolutionary theory is all the greater because the presence of prostaglandins ultimately depends on diet. Muscular activity stimulates their synthesis. Here is a link between expression, physical activity and diet.

The question of gene expression is crucial. However closely the DNA in a cell programmes its structure, other factors also influence that cell's form and function. A key experiment was performed by Gurdon and Uehlinger in 1966. They removed the nuclei from frogs' ova and replaced them with nuclei from cells in frogs' intestines. The embryos developed normally. This striking result shows that what determines how different cells develop into different parts of the animal is not a change in genetic material but is dependent on the genes' environment within the organism. The nucleus from a cell in the nose is just as capable of producing a cell for the eye if it finds itself in the appropriate surroundings. (This is in a sense a true example of plastic heredity because the cells that find themselves directed [by the neural system] into different specialised functions then reproduce cells which perform that same function. Not enough is known about this process but it is clear that 'appropriate surroundings' means conditions, which means physical and organic chemistry.)

Gene expression is concerned not only with differentiation but also with the way in which a cell behaves. As we have seen previously, substrates can alter enzyme activity. The study of gene expression is now providing evidence that substrates can work by acting on the DNA itself. Dr K. Yokuyama in Japan has studied the synthesis of a membrane lipid using simple systems. He identified the specific region in the yeast DNA which coded for an enzyme, choline kinase, that was known to respond to levels of substrates for making membrane lipids. To obtain a clean experiment he isolated this section of the yeast DNA and inserted it into the genetic mechanism of a bacterial strain of *E. coli* which did not possess these enzymes.

This technique enabled him to examine the behaviour of the piece of yeast DNA free from confusing effects of related systems in the yeast. He then proved that the bacterium not only produced choline kinase but that the inserted gene made different amounts of enzyme depending on the amount and type of substrates which he fed to the bacterium. By deleting sections of the DNA code bit by bit, he was able to identify a section of the code which responded to the substrates in the medium and switched on the synthesis of the enzyme. Here is probably the closest proof that external substrates are acting on the genetic mechanism.

Dr Yokuyama also discovered that the substrates that stimulated the gene expression for choline kinase also stimulated the expression of a related enzyme, ethanolamine kinase. Both enzymes are used for membrane growth. Such coincident expression offers a mechanism for the co-ordination of genetic expression.

In Paris Dr M. Mangeney has shown that certain messenger

RNA molecules which translate the DNA information for the cell to make proteins, are under the control of insulin and glucagon, two hormones responsive to the nature of food. Both these types of information describe how external influences can manipulate the amounts of products the cell can make.

John Cairns, Julie Overbaugh and Stephen Millar recently published a paper in *Nature* (vol. 3235: pp. 142–145, 1988) entitled 'The origin of mutants'. They present evidence that bacteria (*E. coli*) grown on a lactose medium respond in a manner which implies that a proportion of the bacteria have mechanisms for making just those mutations that adapted the cell to the presence of lactose which they could use as an energy source. In other words, by plating the *E. coli* which had previously no experience of lactose, on to a lactose containing medium, bacterial mutants arose with the ability to use lactose in a manner which would not have been predicted had there been the odd random lactose utilising bacteria present before the plating. They say, 'The main purpose of our paper is to show how insecure is our belief in the spontaneity (randomness) of most mutations. It seems to be a doctrine that has never been properly put to the test.'

As a mechanism, they suggest a form of reverse transcription leading to a 'directed mutation' which basically means information flowing back from the cell to the DNA. 'If a cell discovered how to make that connection, it might be able to exercise some choice over which mutations to accept and which to reject.' The original studies on such models led people to believe that natural selection was at work. However, in these latest studies it seems as though it is the substrate which is driving the reprogramming of the genetic information.

This type of evidence on adaptive enzymes and directed mutation is only a short step from an understanding of how such manipulations could be ultimately expressed in animal form.

It is perfectly plausible that the change in average height since the turn of the century could have been induced by just such a mechanism. Change in form could be simply brought about by the response of the DNA and enzymes to changing inputs.

Can we use results of this kind to throw light on the central problem of how life systems emerge in new environments, and in particular how they respond to the presence of new nutrients? In the present state of our knowledge any assertion is speculative, whichever side of the debate it takes, so let us speculate. When the distant ancestors of the cat were on earth we can suppose them to have been small animals of a comparatively unspecialised kind, searching for food in an opportunistic way. When one of them developed a taste for eating other animals, it would swallow in one mouthful its victim's store of vitamin A from beta-carotene. The more animals it ate, the more vitamin A and docosahexaenoic acid it would accumulate, both circulating in its plasma and incorporated in its cells. When one of these meat-addicted pre-cats became a mother it would be able to supply the foetus in its womb with greatly enriched levels of vitamin A and docosahexaenoic acid, encouraging gene expression and giving it the chance to develop its eyes closer to the limit of their potential. After birth, this favoured offspring would be better able to see and catch small animals, and so would supply the foetuses developing in its own womb with even more plentiful nutrients necessary for the visual system.

In this way the strain of pre-cats would carry in their bodies a different balance and concentration of nutrients. The protrusions from the brain, for example, which go to make the complex of the retina, the nerves and the iris, would, during early embryonic development, be bathed in a more stimulating chemistry. The genes being encouraged to express their functions could be co-ordinated simply because of a common relationship with the nutrients.

It is not illogical to think that as the pre-cats evolved they would have done so in ways that enjoyed these conditions – ways that fitted them, if you like, to their own inner environment. They adopted new foods which in fact adapted them.

Let us now look again at the possible mechanism for chemical or substrate-driven modification. There is a master plan of genetic instructions for fish, mammals and the like. As mammals, we have two eyes, a nose, two halves of the brain, ovaries, mammary glands, two legs, arms and so on. How could the separate parts of the master plan be manipulated by chemistry? Previously we saw the text book and firmly established body of evidence on 'adaptive enzymes', that is enzymes whose concentrations respond to the amounts of substrates or nutrients in the food.

We have seen how prostaglandins can modulate gene expression and that their synthesis is diet dependent. Hormones also affect cell behaviour and some, like insulin and the adrenal corticoid hormones, are known to be influenced by long-term nutritional forces.

The quality of a protein, or indeed specific amino-acids, can influence protein synthesis and physical growth (e.g. the use of arginine in the body building craze). Hence we already understand how a change in nutritional conditions can lead to modification in body size and shape.

We know from the experiment with the frog's ovum, that when cells divide and become specialised as eyes or ears, certain DNA instructions are shut off and others accentuated.

While the above changes would be considered as plastic, a very simple mechanism can now be proposed which could lead to permanent genetic change. If the introduction of a new nutrient suppresses its synthesis by the cell, it would have the effect of a suppressor sitting on the DNA: just as in cell specialisation. If the new chemical input becomes permanent, then total deletion of the DNA segment would simply not be noticed. As long as the covered DNA codes do not change, the situation can be reversed simply by removing the new nutrient. On the other hand, if that covered section is rewritten, then new ideas could be expressed in a heritable manner.

This heritable change could have happened on a large scale when living systems began to depend on other living systems to provide essential nutrients such as vitamins. A new dimension in the form of the division between plants and animals was created when one cell first obtained nourishment by eating another. The wide range of preformed complex molecules would no longer need to be made. The door was then opened to a new genetic diversity: an explanation for the sudden emergence of the phyla and the answer to Rattray Taylor's second question on sudden radiation.

Modification of existing species could also be envisaged by this mechanism. Continued suppression of a piece of genetic information would allow a change to occur that would become inherited. Thus two types of substrate-driven influences could be envisaged; (i) a plastic form which involved gene expression; and (ii) a fixed form whereby the space occupied by the suppressed information could be re-used for some other purpose.

The balance between different classes of nutrients could also have been important. A high intake of a particular sort of protein could lead to an increase in body size because certain suppressors on codes dealing with amounts are removed or expressions stimulated. It would be expected that such an effect would be widespread and co-ordinated. Similarly, a deficit of EFAs or some other nutrient would turn switches effecting another range of developmental directions, again with an effect on neurological and hence physical shape.

We have seen evidence for this mechanism in the composition of the milk of different species. The milk of the rhino is full of protein and calcium but has virtually no neural fatty acids. Human milk is the opposite. The internal nutritional biology has become set in two contrasting manners and will continue until it runs out of its chosen nutrients.

An Asian girl from the poorer parts of India, if given food typical of the West, would grow bigger than her sister. We would expect her baby to have a higher birth weight, faster growth and become a bigger adult, than her sister's. Where exactly this release of genetic potential stops no one is sure. Elephants and dinosaurs must at some time have been very small creatures. The Mediterranean was created by Atlantic water flooding in through the Straits of Gibraltar. Microcosms of life were trapped on islands. There is evidence that the elephants on such islands were only tiny. Were they, by chance, a specific subset? Or was it that the limited space and food could only support tiny elephants? American parents are now expecting their children to grow up to be over 6 feet tall.

Whether or not the dolphin, whose foetus has vestigial legs, could grow legs again given the appropriate nutritional input, conditions and time we can only guess. It would be interesting to know if the horse could grow hands again – probably not. If it could, then once the phyla were established, evolution becomes little more than embryological differentiation on a grand scale.

Questions remain. We do not know how oxygen created animals, if that is what it did. Nor do we know if linoleic acid created mammals. Did these substances play a passive or directly active role? The links between oxygen and animals, and linoleic acid and mammals are so striking and so co-ordinated, that it would seem to make more sense to think of an active feedback working between substrates and the genetic information: reverse transcription.

The evolution of the carnivores with their articulated claws and the herbivores with their hoofs, was separately co-ordinated and discontinuous. Gould offers a mechanism for the major leaps in a few words and probably better than anyone else:

> Genetic systems are arranged hierarchically, controllers and master switches often activate large blocks of genes. Small changes in the timing of action for these controllers often translate into major and discontinuous alterations of external form (Gould, 1983).

The thrust of his comment is similar to ours. There can be no doubt about the role of genetics, but there can equally be no doubt that genetics alone cannot direct evolution without the materials. Add the possibility of the genetic switches being set by nutrition and we have a potential explanation for co-ordinated evolution in response to the environment: that is, to Darwin's 'higher' law. As we have implied before, no matter how good the architect's plan, the building cannot be constructed without the materials: planning and availability of materials go together and we now have a mechanism whereby nutrients could at the least release genetic space to modify the genetic scene.

Mutations are necessary but not sufficient. Useful ones probably appear over and over again in inopportune circumstances and do not get selected. Everything has to wait for a potentially opportune mutation to occur at a time, in a place and in circumstances congenial to it. Darwin would surely have agreed with this model.

There is, however, a fundamental deviation from the classical view as it is commonly understood today. Nutrition does change shape, form and function before mutation occurs: the horse, the cat and our own dramatic recent increase in height, act as evidence for this effect. The cat may still have the genetic description of the mechanism to convert linoleic acid to its long chain fatty acids, but it doesn't use it. The dolphin has the genetic tapes for a hand but uses flippers instead. Mutation could rewrite these tapes as it probably has done in the case of the horse's hoof.

The conventional view that the horse achieved its present shape and form through competition to run fast, the slower members being filtered out. We suggest it happened the other way around. Nutrition induced change in form, and genetic change, if and when it occurred, would have operated within the confines already set by nutrition. We now need detailed genetic maps to define the extent of these changes: just what sort of a tape is it that replaces the map for the hand in the horse's genetic code? Or is it still there?

Substrate-driven change – or plastic heredity – seems no less plausible than the supposition that the evolution of the cat proceeded through a series of random changes, each unrelated to the nutritional opportunities of the time. Certainly at every step up the evolutionary ladder vital changes were associated with new demands for nutrients and therefore with their supply. As single-cell organisms were followed by muticellular forms of growing complexity, with their need for vascular and nervous systems and new requirements for nutrients; as mammals developed brains that demanded both long-chain derivatives of the new seed acid as well as the original chloroplast acid; as cats, owls and eagles equipped themselves with eyes far more advanced than those of any herbivore, using specialised nutrients; so at every step from algae to cats we can see how the new advance depended on the presence of the right nutrients, the right chemical environment. Most particularly we see how the same nutrient drive led to the development of the same specialisations, as for example in the eye of the cat and the eye of the eagle. Is this just an example of similar gene expression in response to similar substrates?

The conclusion is that substrates could have driven change either forwards in progressive development, or backwards, in retrogression or degeneration. Such ideas support Darwin's original thinking, that external conditions played an important directive role in evolution as well as natural selection.

Exactly the same principle holds true as we come to explore the origins of what must be for us the most interesting of all animal species: ourselves.

8 Apes, dolphins and men

Among all the arguments mustered by the Victorian opponents of evolution there was one that could be unfailingly relied on to arouse a proper mixture of ribaldry and outrage. The wretched Mr Darwin would have us believe that man, so far from being made by God in his own image, is descended from those ignominious little animals the monkeys. The fact that Mr Darwin didn't say that was seldom allowed to spoil the angry joke.

Man, in fact, as now seems beyond dispute, is descended from an ancestor which he has in common with the great apes (which is what Darwin actually said) and the two lines are thought to have diverged around five or six million years ago. (The monkeys branched off on their own particular paths a long time before that.) However, our knowledge of how man evolved after that is based on a small number of fossilised remains and there are many gaps in the record. The anthropologist David Pilbeam of Yale University is quoted as saying, 'If you brought in a smart scientist from another discipline and showed him the meagre evidence we've got [on hominid evolution] he'd surely say, "Forget it, there isn't enough to go on" (Pilbeam, 1984).'

One of the most important areas of dispute concerns how man achieved his present status as an upright biped with a peculiarly large brain. The conventional view is that he passed through a long period as a tree-dwelling and leaf eating primate at the end of which he came down to the ground to live on the open grasslands of the savannah. Life in the trees is supposed to have promoted the growth of his brain. Up in the branches a sense of smell was less important, but a stereoscopic image from eyes placed in front of the head was a great advantage, and that demanded a sharp increase in computer power to interpret the three-dimensional image, especially as it was in full colour and not just black and white. Then again the need to grasp branches made prehensile hands useful and aided the development of the distinctive human thumb; it can move across to close with any of the other fingers, giving a large increase in dexterity for tool making and powered the increase in brain size. Up in the trees, the theory goes, man became a less specialised animal, less dependent on repetitive, inborn patterns of behaviour and more capable of new and appropriate responses to unexpected situations.

When he came down to the ground again he adapted, by natural selection, to a life of hunting game among the tall grasses. The upright posture he evolved gave a selective advantage in two ways. It allowed him to see his prey by peering over the top of the ground cover and it left his forelimbs free to hold weapons and tools. The need to compete with the carnivores resulted in selection pressure for brain size. A weapon-wielding, tool-making hunter is clearly in a position to benefit from high intelligence and in due course he became the most intelligent animal on earth. He was even able to invent techniques that allowed him to colonise the most distant and inhospitable parts of the planet.

Graham Clarke, Professor of Archaeology at the University of Cambridge, and Stuart Piggott, Professor of Prehistoric Archaeology in the University of Edinburgh, expressed the accepted view on human evolution as follows: 'It was the adoption by certain primates of an exclusively terrestial existence that led them to take the decisive steps which led to the emergence of hominids and ultimately man (Clarke & Piggott, 1965).'

New threads of evidence have been revealed by recent fossil discoveries. For instance, it seems that Lucy (a two-million-year-old female hominid discovered by the younger Leakey) was bipedal, was not a tool maker and predated the genuine tool makers that have been found. If true, it would sink the idea that bipedalism resulted from standing up to wield weapons. On the other hand Pilbeam's point about the scarcity of the fossil evidence needs to be remembered. We don't know if her brother, Jason, used giraffe leg bones as clubs or indeed if Lucy belonged to a peace-loving tribe which made knitwear to trade for meat. Bearing this in mind, Clarke and Piggott can again be quoted: 'One of the most important facts about the primates is that throughout all but the final phase of their history, they spent most of their time in the trees.' A tree-living habitat gave protection from the predators and 'by over-emphasising the need for specialised locomotion on four feet and the necessity for agile grasping, encouraged the survival of a generalised structure of the limbs adapted to prehensile functions, with mobile digits with flattened nails in the place of claws.'

When we begin to look closely at this account it becomes less convincing. For a start it is decidedly Lamarckian. Man the tool and weapon maker had a 'need' for a larger brain to exploit his inventions (or at least to exploit the manual dexterity that would later produce them) and he duly grew one. Or if it was simply a matter of natural selection, why did not other predators build such large brains? Arguably, you need to be cleverer to catch game without weapons than with them, and any carnivore would surely benefit from an increase in its brain power. Would anyone care to walk into the tall grass of the savannah to confront a lion with a bigger brain and a more acute intelligence than his own? Baboons are a favourite food of leopards. Why didn't they or any other savannah species competing with or escaping from the carnivores develop the 'big brain'?

There are other flaws in the conventional account. For instance, the growth of man's intelligence is said to have been favoured by the fact that up in the trees he could develop in safety, out of reach of most predators. It was then subsequently favoured by the fact that down on the savannah he was no longer safe, and had to become cleverer to outwit them – a neat example of having the argument both ways. And suppose bipedalism really did evolve before tool making? There is an argument which says that bipedalism was a tropical adaptation designed to minimise thermal stress due to radiant sunlight! So tool making and the ability to kill animals is not seen by everyone as the reason for upright stance.

The idea of an upright primate scoring by being able to peer over the tops of the grasses is an appealing one – to anyone who has no experience of hunting. In reality the main difficulty facing any hunter is not spotting his prey but preventing his prey from spotting him. The art lies in the stalking. A polar bear, for all its white camouflage, will slip into the icy water and move, virtually submerged, towards the seal lying on the edge of the ice. If you watch a cat stalk a bird, it squeezes its body as close as possible to the ground: it seems to flow forward in controlled, silent motion with its eyes fixed unwaveringly on the position of its prey. The big cats do the same with frightening ease.

Darwin's own account of the Aborigines also contradicts the idea of upright stance as a byproduct of hunting. He was undoubtedly an accurate observer and in his book *A Naturalist's Voyage Round the World* he described the attempt to drive the aborigines out of Tasmania.

> A line was formed reaching across the island, with the intention of driving the natives into a cul de sac on Tasman's peninsula. The attempt failed . . . the natives stole one night through the lines. This is far from surprising, when their practised senses and usual manner of crawling after wild animals is considered . . . I have been assured that they can conceal themselves on almost bare ground (Darwin, 1845).

A hunter stalking antelopes or wild pigs with a modern rifle will do his best to emulate the Tasmanian and the cat. Creeping about the savannah on your stomach is extremely uncomfortable but unless you want your target to spot you first it is what you had better do, even if it means that for much of the time you cannot see the animal you are stalking. Beginners who attempt this method, or still worse try to move crouching on all fours, often betray their presence by their give-away rear end protruding above the grass. Anyone trying it will soon be left in no doubt that the human anatomy, with its upright stance, is not designed for stalking prey.

Where an upright posture is an advantage is in throwing a spear or wielding a club, but it is possible to exaggerate the degree of adaptation this demands. Other primates can stand on their hind legs well enough to free their hands for particular jobs. Baboons will raid crocodile nests on the banks of the Nile and can be seen running away on two legs with armfuls of eggs. Many of the eggs get dropped and broken, but that does not alter the fact that the baboon can move fast on its hind legs only. Quite a few primates resort to standing or walking on two legs when they need to, even if they try not to make a habit of it. Squirrels often stand upright when nibbling at nuts held in both hands.

The real requirements for throwing a missile are a sense of balance, and a versatile arm. Balance is largely a function of the cerebellum, a primitive part of the brain which is common to many animals, well developed in birds and used in flight – whereas in man it is the cortex and frontal lobes which are outstanding. The evolution of the versatile arm and shoulder joint is not a specific feature of life on the savannah because it is shared with other primates: of the four-legged animals it is only the top carnivores that can also move their front limbs sideways. The front legs of dogs, deer, horses and cattle can only move backwards and forwards with a little sideways play: domestic cats will 'box' or knock a ball of wool around with their front paws, and a lion will kill an animal with a cross swipe which is powerful enough to break the human neck.

Developing an upright posture presented the human animal with serious problems, not all of which have been solved even now. The obstetrician Professor Donald Gebbie comments on the obstetrical incompatibility of upright stance and a big brain. A bipedal mother really requires a small pelvis whereas the large brain requires a large one. During early hominid evolution the pelvis had to develop a new shape and new muscles: 'To prevent their intestines from prolapsing to the ground between their legs, a pelvic basin became necessary, formed by an extension of what is known as the "false pelvis". Muscles which in four-legged species were used for the "genial pastime of wagging the tail" had to be employed in "more mundane duties such as containing the internal genitalia and the rectum in their places" (Gebbie, 1981).

Gebbie also makes the important point that the female pelvis had to develop a large enough passage to allow the head of a big-brained baby to pass through. Modern women often suffer from disproportion here, resulting in painful labour and sometimes necessitating Caesarean section. Most people explain this as a result of our increasing brain size, but Gebbie produces an answer which, once stated, seems obvious. The problem is not the brain but the pelvis. The upright posture demanded a pelvis strong enough to allow women to walk and run, but not so capacious that it let everything fall through. The modern birth canal is a risky compromise which usually, though not always, leaves just enough room for the head of the emerging child. Gebbie's experience from working in many different countries led him to the view that it is poor skeletal nutrition which gives rise to a poor pelvic dimension. He sees pelvic nutrition as the more salient explanation of disproportion rather than the size of the brain case.

Once again, when we try to discover how the human body arrived at its present proportion, we are forced to rely on the fragmentary fossil record which provides us with only a few specimens to cover millions of years of evolution. One of the first erect human-like species we know of was *Australopithecus*, and Gebbie comments on

Table 2 Some brain/body weight ratios in average adults of various species

	Gestation length (days)	*brain weight* (kg)	*body weight* (kg)	*brain weight as % of body weight*
ox	280	0.45	520	0.090
hippo	240	0.59	1400	0.042
elephant	720	4.00	3000	0.130
chimpanzee	265	0.34	62	0.550
gorilla	230	0.406	1400	0.240
human	280	1.500	70	2.1
squirrel		0.020	0.890	2.2
dolphin		1.600	155	1.0

Note: The large savannah animals have brain capacities which are small by comparison with small mammals, marine mammals and man. Presumably there must be a lower limit to brain capacity, below which the capacity is inadequate to fully monitor and control the functions of the body which it serves.

a fact about this hominid which is curious. The brain capacity of *Australopithecus* was 700 ml, which is twice that of the chimpanzee but only about half that of a modern human brain with a volume of 1300–1500 ml. The birth canal of the female *Australopithecus* however, was big enough to accommodate a modern skull. It seems that she had a much larger birth canal than she needed, and since this would have had serious disadvantages for the upright stance there must have been a reason. But what? We seem to be left with two alternatives. Either she showed remarkable foresight in developing a modern-sized birth canal millions of years before it was needed or things happened the other way round. The discrepancy is explained if we assume that the relative size of the birth canal remained the same as the relative brain size shrank.

From what we know about present-day land mammals this seems possible. A cow's, lion's or even a gorilla's brain is considerably smaller than a modern human's in relation to body size. The horse's brain accounts for about 0.02 per cent of its total weight. The size of the human brain relative to its body is 2 per cent.

The biologist Bill Pirie has on several occasions made an important give-away comment which is this: the squirrel's brain is, relatively speaking, as big if not bigger than our own! Its brain/body weight ratio is 2.2 per cent. Indeed, the size of the brain relative to the size of the body in small mammals is much the same but the brains of the savannah species, at a brain/body weight ratio of less than 0.1 per cent are tiny. This of course fits with the predictions, because the savannah species are very large, but it is odd that as these animals became bigger, their brains became relatively smaller.

The classical interpretation of evolution is of 'improvements' step by step through selective advantage. What possible advantage can it have been for a chimpanzee or a lion to lose relative brain capacity to such a massive extent?

The classical view sees *Homo sapiens* as evolving a big brain from a little one. The typical image was portrayed in the 'just Genius' advertisement for the black beer Guinness. The picture depicted a row of apes, starting with a stooped, virtually four-legged version of the pin headed chimp. The next figure was slightly more upright, but still with huge brow ridges and sloping cranium clearly indicating not much inside. The further figures were progressively more upright in stance with the bigger and bigger brains, and finally bowler hatted *Homo sapiens* holding a glass of Genius.

Now supposing that story was false. Supposing it was the other way round. As the lions and gorillas became bigger, their brains became smaller in relation to the size of the body and the basic physiological demands it imposed. Then the line that became *Homo sapiens* did not need to evolve a bigger brain at all. All he did was to *keep* what he had as a small mammal.

He simply kept his brain when all around were losing theirs. When one recognises that the savannah species 'lost' relative brain capacity and man retained it, then one can view the origin of man's brain afresh. Starting with the large pelvis and small brain of *Australopithecus*, more sense can now be made of the history.*

There is no certainty that the Australopithecines were direct ancestors of man; indeed there is considerable dispute about it. Dr L. S. B. Leakey considered: 'It is reasonably certain that both the Australopithecines and Pithecanthropines represent no more than

* There is a view that brain size is related to body size in a logarithmic manner. This means that with bigger bodysizes there is less brain – and that's fine because that does the job. However, *Homo sapiens* is exceptional and doesn't fit these predictions. According to these calculations the volume of our brains should be about a half-pint size: whereas it is actually nearer 3 pints. It is three or four times bigger than it should be. How did this happen?

side branches and are not ancestral stock to *Homo sapiens* (Leakey, 1970).' Using the evidence on brain capacity, we would tend to agree and suggest that this species, having become land-locked, became a degenerate side-shoot which, like the savannah species, followed the common path of increasing body size outpacing the growth of the brain.

This pattern holds even for the primates; intelligent as they undoubtedly are, they are still far behind us. The chimpanzee, with a brain weight of around 300 to 350g and a body weight around 60kg, has a brain/body ratio of about 0.55 per cent, compared with the human 2 per cent. The gorilla's ratio is only about 0.29 per cent – brain weight around 400g, body weight around 140kg. There is simply no evidence of any of the savannah or forest species doing anything other than losing out in terms of brain capacity and the allometric calculations simply confirm this conclusion.

Did we lose relative brain capacity and become like a chimpanzee, then grow it back again? Or did we simply never lose it in the first place?

Relative brain size is not a measure of intelligence. It is likely that the number of brain cells and synaptic connections is the key. Also, there are certain economies of scale, so that a larger animal could take a small drop in relative brain size without becoming less clever; but this is a comparatively minor effect and in no way contradicts the general rule that mammals that grow larger end up with a smaller usable brain capacity. Every part of the body tells the brain where it is, what it feels and what to do. This intimate relationship with fingertips, lungs, bones, intestines, feet, tongue, heart and eyes, communicates a constant flow of millions of bits of information, the bulk of which we are totally unaware of until something important happens. The brain and nervous system process information, make decisions and take action, often without reference to the conscious, especially if dealing with internal information. Inputs from outside involve the senses of hearing, vision, touch, smell and taste; they only invoke conscious decisions if the monitoring system detects some action which demands attention at a higher level.

Hence there is a constant monitoring of the affairs of the body which require priority attention from the central processing unit in the head. There will be a basic size of central processing unit (CPU) needed to cope with these mundane matters. A huge body will demand more attention than a small one and the increase may not just be linear because many of the increments will involve surface areas (e.g. skin, bones and intestines).

Whilst the brain of a rhinoceros is many times bigger than that of a squirrel, a domestic cat or an otter, one suspects that these animals have more spare capacity for games and a generally more complex behaviour pattern. By contrast, what little brain the rhino has is largely tied up in mundane business.

In the tragedy of Alzheimer's disease, the brain shrinks and with it goes intellectual ability, memory and that sum of behaviour which makes a person. All that is left is basic monitoring with only occasional flashes of identity. The person is deprived of the higher functions that the rhino never had.

Why is man's brain size the exception? There are no species in the forests or the savannahs exhibiting any development of the brain that paralleled *Homo sapiens*. Is there any other place where it occurred? The only other large species to remotely approach man in relation to brain capacity is not a land mammal but a marine mammal: the dolphin. The adult dolphin's brain runs to about 1600g, slightly larger than the human brain and giving a brain/body ratio of between 1.0 and 1.5 per cent – uncomfortably close to our own. Is there a connection? Do we and the cetaceans have something in common that could account for this similarity?

In an article in *New Scientist* in 1960 Sir Alister Hardy suggested that man's final evolution after he came down from the trees took place not on the savannah, but in the sea. Hardy's idea attracted the attention of Elaine Morgan who wrote a book in 1982 called *The Aquatic Ape* which elaborated on the thesis. A branch of the line of primitive ancestral apes was forced by competition to leave the trees and feed on the seashores. Searching for oysters, mussels, crabs, crayfish and so on they would have spent much of their time in the water, and an upright posture would have come naturally. It would also have been favoured by two factors: first, the reduced weight of their bodies in water would make the upright position easier to sustain, and second, when they learned to swim (as they surely would) they would be in a flat, extended position.

This revolutionary idea was met with ridicule: some people even said that Hardy's tongue was firmly in his cheek as he had been asked to give the lecture to a sub-aqua club. We believe that this proposal should be taken seriously, though we propose an important amendment. In our view the main primate line leading to modern man did not come down from the trees to enter the water: that would have involved losing his brain and clawing it back again. More probably, he never went up into the trees in the first place. The primates that did became degenerates.

Five or more million years ago the dolphins were common land mammals discovering the attraction of sea foods, and, like many other land mammals, they were taking the decision to leave the land and recolonise the sea. Indeed a group of mammals which had once been on land migrated into the sea and became the marine mammals. Some, like the seals, never wholly left the land: they frequently come out of the water and indeed do so to give birth to their young. But their food now comes from the sea. At the other extreme, the savannah species were exploiting the land-locked regions and became adapted to the hot dry environments on offer. This scenario leaves an intermediate niche in the ecosystem untapped: the land–water interface. It is highly unlikely that nature would leave such rich food resources unused. The desert rat explores the extremes of aridity, and the snow leopard the high altitude snow line. Why would biology explore every conceivable and often poverty stricken food resource, and leave what is perhaps the richest untapped?

It makes more sense to say that, like the dolphins, one branch of the hominids found that the sea offered a wealth of food and a way of life that was congenial, much in the way that we enjoy the seaside today. This species would have taken to the shores of the freshwater lakes and rivers as well, and the adoption of an aquatic habitat would not have cut them off from other supplies of food, for the coastal regions, with their high humidities and ample rainfalls, offer equable climates and a rich growth of fruits and other vegetation. The estuaries would have been sites of particular value because it is here that the marine food chain begins in earnest, fertilised by the trace elements and minerals washed off the land, and it is here that early man could have had an abundance of food and fresh water.

If man never went up trees, and there is no good reason why he should, there is also no reason to believe that he was wholly aquatic. He most probably evolved at the land–water interface. At different times it is likely that he would have explored inland regions, and in this way various side shoots may have developed with some, like *Australopithecus*, suffering as a result. What is certain, is that nowhere in the world today is there any evidence of modern man having made any adaptations to the semi-arid savannahs. It is worth remembering that most of the animal phyla are actually water-based: land-based animal life represents the exception rather than the rule.

For example, recent classifications now recognise about 30 animal phyla: three are parasitic; 11 are specifically marine phyla; and 16 are represented in sea or fresh water. Only one phylum out of thirty is claimed to be totally restricted to land. Hence the movement of mammals, backwards and forwards between land and sea, should not be too surprising.

Sir Alister Hardy pointed to a number of characteristics which are unique and out of place in a land animal but which are quite appropriate in the water and which we share with the aquatic mammals. First he comments on man's exceptional ability to swim and his endurance in the water, which can stretch as far as swimming 22 miles across the English Channel and even back again. Many mammals can swim to some degree but few, Hardy remarks, can rival man in swimming below the surface and navigating underwater.

Darwin again provides our evidence. When he visited Tahiti on board HMS *Beagle* he noticed that the Tahitians had 'the dexterity of amphibious animals in the water'. When they acted as Darwin's guides on expeditions to the interior, they did not carry food but instead took a small net stretched on a hoop. 'Where the water [of the river] was deep . . . they dived, and, like otters with their eyes open, followed the fish into holes and corners and thus caught them.' Polynesian pearl and sponge divers can reach depths of 175 feet by simply holding their breath and using a stone to carry them down. Impossible? Or peculiar to the Polynesian races? No; the fact that Turkish sponge divers can also dive to great depths led them to the discovery in 1982 of a pile of gold ingots on the sea bed from the oldest complete shipwreck ever found. The ship was wrecked about 3400 years ago and lies at a depth of 150 feet at Ulu Burun, near Kas. The wreck contained tablets which, according to Dr George Bass of the Texas Institute of Nautical Archaeology, described scenes of gifts almost as though they were shopping catalogues. They referred to goods available at a time just before Tutankhamun.

The objectors to Hardy's theory point out that because we have to learn to swim, it cannot be said to be natural to us; but in fact the same thing is true of fully aquatic mammals. Young otters are taught to swim, and even the baby dolphin, is helped and guided by its aunts immediately after it is born. It is also true that if a human infant is put in water before it is six months old it will swim spontaneously with its eyes wide open and breathing controlled. Later it loses its innate ability: unless started at this early age, it has to learn to swim the hard way.

Hardy also refers to our loss of hair, which is characteristic of a number of marine animals. The exceptions, which have kept their fur, are those that do not spend all their time in water but, like the seal and the otter, come out on land in cold climates where they need the fur to keep warm.

Linked to hairlessness there is a series of glands that produce skin secretions known as the eccrine glands. In general, the more hair an animal has the fewer its eccrine glands, and it is significant that while monkeys and apes have very few, such water-going animals as the hippo have a large number – and so does man. Furthermore, man has a layer of fat under the skin, like the marine mammals but unlike the other primates or the savannah animals in general. Richard Dawkins, during a BBC interview on his book *The Blind Watchmaker* said he thought man became hairless because men liked to copulate with hairless women and so selection gradually produced hairless offspring. Curiously he did not suggest the opposite and one wonders why. The Uganda kob is territorial, stands on his patch on open savannah and at the peak of his prowess inseminates 32 females per day. If this is the law of the jungle then a far more effective technique would have been for the women to prefer copulation with hairless men: a *reductio ad absurdum* or just plain chauvinism?

An obvious and very significant point concerns the way we control our temperature. Most species of the African savannah have evolved techniques for dealing with the heat without losing too much water, like the oryx and eland. Those which have not, like the buffalo, the water buck or the sitatunga (the marsh antelope), are always found close to water. Perspiration is not a technique the savannah animals employ: it is too expensive, except for an animal like the hippopotamus which does sweat but makes up for it by sitting in water or mud during the heat of the day and feeding only at night.

Michael Crawford recalls: 'I discovered just how important this is to the hippo on an expedition to Uganda with Professor C. P. Luck of the Makerere Medical School. One of Professor Luck's research interests was temperature regulation in animals and in pursuit of this he used a kind of hypodermic syringe shot from a crossbow to anaesthetise anything from small antelopes to giraffes or elephants. By this means he was able, for example, to discover why African elephants stand and flap their ears. The blood leaving the ear and returning to the body is noticeably lower in temperature than when it first enters the ear, which has an extensive network of capillary blood vessels and makes an effective cooling device. (The Indian elephant, which spends its time in the shade of the forest, has by comparison small pig-like ears.)

'One hippopotamus which Professor Luck anaesthetised just as it was about to return to the water lay for no more than five minutes on its stomach in the early morning sun. On the equator it gets hot very quickly and in that time the hippo's whole body became bathed in a thick layer of perspiration which oozed in visible motion out of every pore. Indeed the pore holes were so large that slender plastic tubes could be inserted into them to measure the flow. It was simple to calculate that the animal could not sustain the rate of water loss that we were witnessing for more than an hour, so once the necessary scientific observations had been made and specimens collected, the hippo was given an antidote and the human crew retired to the safety of the Landrover. The hippo trotted off to his pool, into which he sank with a shake of his head and some rather violent bubble-blowing mingled with a series of deep-throated snorts.

'We had parked the Landrover on a convenient track that led through the bushes to the hippo wallow. While our attention was focused on the animal now enjoying it, a second hippo had been humping down the track behind us. Hippos tread softly on the ground and it was not until he was a few yards away that we noticed him. Hippos are not normally aggressive beasts: they usually mind their own business. To our surprise, instead of doing the sensible thing and overtaking in the fast lane, this hippo decided simply to accelerate towards us. Professor Luck pushed the starter button, the engine sprang to life, the vehicle leapt forward – and stalled. At that moment we recognised that we really ought to be somewhere else.

'It seemed that the hippo meant to demolish the blockage by simply eating it. The cavernous mouth opened and four tusk-like teeth sank into the back corner of the Landrover, where a moment before I had been sitting. A second try with the starter got the Landrover struggling forward with a hippo clutching its rear end. Fortunately it let go, perhaps deciding that aluminium did not taste very good. But it had given us a crystal clear demonstration of how serious and single-minded a hippo is in getting to his watery wallow on time.

'The point, of course, was that we had not just parked on any old track but on his very own track which he had made over the years. It was his path to the water. The Professor was hugely proud of the gaping holes in the back of the departmental Landrover and studiously avoided having it repaired until the next rainy season forced his hand.

'In savannah conditions man too is helplessly dependent on a supply of water, though he only needs to drink it, not wallow in it. On another expedition in Uganda with Neil Casperd we measured the loss of water from our bodies when we were on a foot safari in Tonia-Kaiso. We discovered that between 10 a.m. and 4 p.m. we were losing it at the rate of 1.6 litres an hour, or over two gallons in six hours. Our African colleagues were losing water at the same rate.'

Recently some French skiers thought it a good idea to see if desert sand dunes would behave like off-piste powder snow. They each drank two gallons in an afternoon. Our physiology does not make sense in a savannah habitat. Even if standing upright does minimise the effect of solar radiation, the upright posture still leaves us hopelessly vulnerable without water.

Animals in an arid environment cannot possibly afford to treat a very scarce resource with extravagance. If man is a savannah species he is the only one that rigidly controls its body temperature and uses a copious loss of water through the skin to do so. If, on the other hand, man has an aquatic background such a physiology would be quite expected. In addition, human babies, like baby dolphins and unlike baby chimps, are born with 300–400g fat: we do not use that fat like a camel to provide water but it would be very useful for buoyancy. Again we have a layer of fat under the skin which we have in common with the marine mammals but not the chimps or the savannah species.

Another astonishing discrepancy pointed out by Hardy is our diving reflex. When seals, dolphins and whales dive, their heartbeat slows down to 10 to 12 beats a minute, the pulse slows down and their blood pressure drops. By this means they can conserve oxygen, which during a prolonged dive is obtained chemically from a compound known as myoglobin found in the muscles. There is now a lot of data on this, obtained by fitting captive animals with radiotelemetric devices which transmit information on physiological functions in the same way that an astronaut's heart beat and blood pressure are monitored in space. A seal, frightened into taking a deep forced dive, can slow down its heartbeats until they are as much as 60 seconds apart.

That the human species possesses this diving reflex is only astonishing if man had a savannah origin. It is only necessary to immerse someone's head in water and his heart rate slows and blood pressure drops; indeed it is so predictable that it has been demonstrated live on UK television. It is one more of those attributes of man which are not found, and are certainly not what you would expect to find, in savannah species. A diving reflex would be expected against an aquatic background.

Another human characteristic, out of place in a land animal, is our extraordinary and persistent love of water. There seems to be some deep affinity which persists through age after age of history and across all the barriers of culture. Water both excites and soothes us like no other element. We wash in it, dive into it, use it as a playground wherever it is warm enough to do so (and often in places where it is not). Children, like hippos, put their heads under it and blow bubbles and, despite our Victorian inhibitions, the greatest pleasure is to swim naked in it.

A visit to Ephesus illustrates just how important the bath was to the Romans. They made the public bath one of the centres of their society. They sculpted boys riding on dolphins; perhaps not such a curiosity when it is remembered that Romans used dolphins to round up fish.

Still today, from the Welsh hill spa of Llandrindod Wells to the hot springs of Rotorua in New Zealand, people soak in mineral waters to calm their minds and cure their bodies. In Bad Bellingen, a few kilometres north of the Autobahn from Basel to Strasbourg, is a beautiful new cultural centre built in 1983 around a hot sulphur spring with an indoor and outdoor bath section, evidence of the stable popularity of bathing and drinking mineral-rich waters in the interests of health and well-being.

In wealthy countries almost every house has its hot bath or shower, used for relaxation and pleasure as much as for keeping clean. Public swimming-pools, jacuzzis, every crowded mile of the beaches of the Mediterranean, Mombasa, Florida, California, the Black Sea and Bondi, all proclaim mankind's kinship with water. Even at a late stage in his evolution man loved the water and used it to explore the planet. Water enters time and again into art and literature, and from Christian baptism to the Hindu's immersion in the Ganges, water is a symbol of profound religious importance.

Obstetricians Michel Odent in France and Igor Tjarkovsky in Russia believe that it is easier and better for babies to be born underwater. The release from gravity and the soothing nature of the aquatic environment are claimed to contribute to ease and comfort. The baby swims out from the water of the womb to the same, familiar environment in the outside world and is gently introduced to the reality of gravity and air. Tjarkovsky created a sensation with his photographs of newborns swimming and diving like baby dolphins (Sidenbladh, 1983). He became interested in dolphin behaviour and studied it at the Black Sea Dolphin Research Station. From his research, he is convinced that the extraordinary way in which dolphins care for their own young is expressed when confronted with human infants or very young children. What is more, the feeling is reciprocated. Newborn humans immediately achieve some form of understanding.

Dolphins, although carnivorous, do not usually attack people. Tjarkovsky records one occasion when two dolphins alarmed people who were playing in the water by swimming in from some distance at great speed and heading for a group of children. The dolphins must have heard the distress calls of a boy who was in fact drowning, for they lifted the child to the surface. No one else had noticed. Dolphin aunts behave as midwives at a birth, helping the newborn to swim, taking it to the surface to breathe air and even nudging it into the right position to suck its mother's milk. Surprising as it may seem, Russian researchers observed one of their colleagues giving birth underwater when the dolphins wanted to behave as though it was the birth of one of their own kind. Unless one has actually seen this kind of behaviour, it is so astonishing that it is difficult to believe.

The idea that man passed through a phase of aquatic life should not be too surprising. There have been many examples in evolutionary history of land plants and animals moving partly or wholly into the water, from the water-lily and the ichthyosaurus to the beaver and whale. There is no obvious reason why a species should not enjoy both water and land, which is precisely what man does universally to this day. The majority of the fossil remains of early man have been found inland, but that is not surprising. We have few, if any, fossils of intermediate dolphins as they evolved from land to marine forms, and the reason is obvious. The seashore is in constant flux, being eroded, changing shape and shifting place. It is very unlikely that a fossil will endure for millions of years in such conditions, whereas in a dry place inland it is far more probable.

We must not exaggerate the extent to which man, even today, is a land-dwelling animal. There is plenty of evidence that the oldest human settlements were at the margins of water and the great cities of the world are close to the sea, rivers or lakes.* The earliest cultures of Europe and the Near East were built around water, in Mesopotamia between the great rivers Tigris and Euphrates, in Egypt along the Nile, on Minoan Crete in the Mediterranean. The five civilisations to evolve written languages all emerged on rivers, or estuaries; the Yangtse, the Ganges, the Indus, the Nile, the Tigris and the Euphrates.

Some historians and social anthropologists argue that cultures emerged beside water because it allowed them to trade. Throughout most of history it was easier to travel on water than on land: the sea linked the Mediterranean cultures, the difficulties of land travel divided them from the peoples in the land masses behind. However, we need again to analyse this aspect by starting from the beginning. Sean McGrail, Professor of Maritime Archaeology at the University

* Just some of the main cities by the sea are: Abu Dhabi, Alexandria, Amsterdam, Athens, Auckland, Bahia Blanca, Bombay, Cairo, Calcutta, Casablanca, Copenhagen, Dar-es-Salam, Edinburgh, Gdansk, Genoa, Glasgow, Helsinki, Hania, Ho Chi Minh, Hong Kong, Istanbul, Kuala Lumpur, Leningrad, Lisbon, London, Marseille, Melbourne, Oslo, Rangoon, Rome, Sebastapol, Seoul, Sidney, Singapore, Stockholm, Tokyo, Vancouver and Venice. Many others like Berne, Geneva, Khartoum, Kampala, Paris and Vienna have grown up around inland lakes and rivers.

of Oxford, made the point that the great maritime migrations could not have happened unless by that time man was already familiar with water, fearless of it and able to swim in it. Primitive peoples did not start trading by water, they first found they enjoyed messing about on logs, then began to use them for fishing long before they ever thought of building boats, never mind using them to travel long distances. Man would never have taken off on the great transmaritime migrations had he not first been able to swim and feel comfortable in water.

Man probably first learnt to swim when he was no bigger than an otter. It is an obvious statement of the necessary chronology that he must have been swimming and fishing and developing social groupings at the water's edge long before sea trade would occur to him. Indeed, the development of tools might just as easily have arisen from his desire to make rafts, nets and then boats. It was the intimate contact with water that explains cultures emerging beside water. Yet again Darwin's observations provide us with further evidence of *Homo sapiens* living by water. As they approached Tierra del Fuego in his *Naturalists' Voyage Around the World* in 1845 he wrote that the *Beagle* had at night 'endeavoured in vain to find an uninhabited cove'.

Trade unquestionably became very important, but that does not mean it was the original motive for settling at the edge of water. We can see this demonstrated in the history of a people whom Europeans have watched from their 'stone age' onwards – the Australian Aborigines. Forty to fifty thousand years before the Europeans arrived, the Aborigines had established their peaceful culture, which regulated their population and lived as part of nature, treating the earth as their mother. That all changed in 1786 when Captain Cook sailed between the Spanish and French ships into what is now Sydney Harbour and raised the British flag with a nod of thanks to their surprised commanders standing timidly offshore. He and those who came later found the Aborigines living along the coastal regions, combining the food produced by the land with that of the rivers and the sea. They had not developed a settled agriculture or animal husbandry because they had no need. Food was there for the taking, and though their walkabouts took them far across the great land mass, the rivers and seas were rich providers of food.

The English drove them out of their best sites along the coast and the rivers and forced them to exist in impoverished land. Darwin visited Australia on his voyage round the world half a century later and reported: 'All the Aborigines have been removed to an island in Bass's Straits so that Van Diemen's Land enjoys the advantage of being free from a native population.'* The British colonisers were certainly moved by a wish to expand their trade, but we should not read that motive into the cultures they displaced. There is no evidence that the Aborigines had traded with anyone by ship. There is evidence of the mixed diet from land and water they enjoyed. For example, they developed four main types of spear: for fishing, for hunting turtles and dugongs on the coast, and for killing kangaroos, wallabies, lizards and other land animals.

Professor Donald Gebbie considers that the Aborigines had ample, excellent food and little disease as a result. He argues that their society wove into its behaviour a relationship between food and population control so that having already survived for 50,000 years they could go on for another. Until the European came, they had solved the man–food equation. Alcohol and purified carbohydrate foods, and poor soil were no substitute for the diversity to which they had been accustomed. Today, despite pollution and the other destructive impacts of European culture, there is still ample evidence of the wealth of the marine environment as a source of food: for the less demanding needs of the Aborigines there would have been a more than adequate supply.

The same holds of the Polynesians. The islands of New Zealand are the southernmost in the great chain of Polynesian islands strung across the Pacific, and were the last to be discovered by the Polynesian people, who reached them only about 1000 years before the British. There is no doubt that the Polynesians were great sailors but that does not make them traders. Sir Peter Buck, the Maori historian, tells us that they left home to discover a new land to resolve frictions – family feuds and the like. They did not go to find new trading centres or even natural resources: they had all they could want in the sea. And when they reached New Zealand they did not use the sea routes to trade with the Aborigines of Australia, although they would certainly have been capable of it.

The Vikings, like the Polynesians, were a great sea-going people

* The phrase is a sad give-away. Darwin does not seem to have been troubled by thoughts that the Europeans' fitness to survive where the Aborigines went down was a function not of their superior culture but merely of their improved technology for killing.

who made long and adventurous voyages, discovering Iceland and Greenland and even reaching Canada at roughly the same time as the Maoris settled in New Zealand. What the familiar picture of Vikings as adventurers and warriors omits is how they became set on that course in the first place. No culture builds its first boats as weapons of war. It builds them to get on to rivers or on to the sea close to the shore in search of that first requirement of all, food. Once they did travel, far from engaging in trade, their first reaction on meeting the British was to rape and plunder.

We may be certain that before the Vikings took to voyaging, plundering or trading, their ancestors built boats, probably starting from two logs tied together, to take them out fishing. The same must have been true of all sea-going peoples. Long before they could learn the art of navigation they had to discover how to make boats, for which the making of tools and ropes was a prerequisite.

Living beside water had to come first and what drew them to the water's edge was not the distant prospect of trade in some dim future millennium, but the immediate access to food. When they had settled down inland the Vikings continued to depend on fish for a large part of their diet. Excavations at various places including the Viking city of Yorvik (present day York in the north of England) have unearthed evidence such as sea shells attesting to the prominent part that seafood played in their diet just as it does for Scandinavians today. Throughout the world the most land-locked peoples eat fish when they can, and they have done so throughout history and prehistory.

We think of Britain as a meat-eating country yet history would seem to suggest otherwise. Sir Walter Raleigh persuaded Queen Elizabeth I to fight for Newfoundland for no other reason than the rich cod fisheries. One half of England's income was at that time dependent on fish. Mrs Beeton's cookery book, that great repository of traditional English recipes, also witnesses differently. Out of 340 recipes for main meals no less than 280 include fish or other products of fresh and salt water. Even the breakfast recipes, far from being just a matter of bacon, eggs and sausages, are often based on fish and other seafoods. The traditional 'Cockney' food of Londoners was cockles, mussels, whelks, oysters, scallops, crabs and a variety of fish from the Thames estuary. Although England now relies on supermarkets and much land-based food, this reliance is recent and a result of rapid population growth and urbanisation.

Fourteen million people live within a 20-mile radius of central London and the Thames and its estuary is polluted by them and their industries: not much hope of wild foods there! No more than a century ago it was different. The proud claim on shop hoardings and the salesman's cry was, 'Mussels and cockles alive, alive, o!' to indicate freshness.

Similar foods were being eaten by the natives on the shores of South America when Darwin visited it. Seeing the piles of shells, he expressed his dismay and pity for those who had to live on such 'miserable food' – an odd description, perhaps, for lobsters, giant prawns, oysters, abalone, scampi and mussels. Data from the Food and Agriculture Organization in 1980 shows that world-wide man uses the same amount of fish protein as he does animal protein! This 1:1 balance is only astonishing because so much has changed in Western culture in recent decades.

Even in Central Africa, dependence on fish as food goes far back. Michael Crawford remembers: 'I had the pleasure of accompanying Professor D. Albrook, the Professor of Anatomy at Makerere Medical School, on a tour with Professor J. Z. Young, the eminent zoologist who was visiting Makerere as an external examiner, we went to what was then called the Queen Elizabeth National Park in Uganda and spent our time looking for archaeological remains close to the shore of Lake Edward. We had the extraordinary good fortune to come across a small burial ground, about 5,000 years old. Next to where a body had been laid to rest, in a foetal position, we found pottery, some still almost intact. Inside were remains of charcoal and fish bones. It was hugely impressive when Professor Young identified the species of fish from the bones in the pots. We were also surprised that what we found were fish bones, not animal bones although we were in one of the richest game regions of Africa.'

The same thing was discovered from excavations of the Kastelberg site on the Vredenburg peninsula in the south-western Cape Province of South Africa. Gas-liquid chromatography was used to examine the fats found in pottery fragments, and it seemed that a minor part came from sheep and antelope but the major part marine animals and birds.

The pleasure of fishing, which grips so many people today like a passion, clearly has a long ancestry. It is indeed, ritualised in many countries. People sit for hours beside a lake, pond or canal and actually compete to see how many fish they can catch: they put the fish back in the water! Some consider this most odd; is it a subconscious throwback to our evolutionary history?

Whilst one culture eats wheat and another rice and yet another sweet potatoes, all cultures across the world have one food in common: seafood. During the Vietnam war, the North Vietnamese turned American-made bomb craters into fish ponds! The Chinese today occupy great tracts of inland continent but they grow rice in watery paddy fields and in them they farm fish. The Russians, who appear to be largely land-locked, have one of the largest fishing fleets in the world. They and the Japanese were the most intransigent over sparing the blue whale. The Japanese make the greatest use of fish of any nation. Indeed they could be said to be predominantly a fish and seafood nation. If the Japanese were to grow enough food on land to replace the amount of fish they eat, they would need seven times the amount of land which they presently possess.

We have, then, four facts to deal with: (i) we humans have a close connection with water which includes a continuing reliance on water-borne foods; (ii) anatomically and physiologically, we have features in common with the aquatic mammals that other, purely land-going species do not possess; (iii) we alone among land mammals have retained our relative brain size as our bodies evolved to their present size; (iv) the only other species to have done the same is aquatic. Can we put these facts together? Let us first look again at the human brain.

What distinguishes the human brain from the brains of other mammals – not merely cows and horses but chimpanzees and gorillas as well – is not its size alone but its structure. We are not in all ways better equipped than other mammals, particularly the carnivores. The dog's nose is incomparably more sensitive than ours, the owl's night vision much more acute. The bat uses a form of hearing far beyond our capability to catch insects in the dark at full speed. All of these senses depend on highly developed receptors but they also demand regions in the brain to deal with the information the receptors send in. We have to accept that in those regions the bat, the owl and the dog have better brains than ours.

What we have, and they have not, is a large volume of cerebral cortex, the 'grey matter' that forms the outer layer of much of the human brain. This cortex is the part of the brain that is taken up with dealing with immediate sensory inputs and motor outputs. It is free to do its own work, and that work is to provide us with intelligence. It is like the top management of a well-run company which, because it does not need to spend its time looking over the lathe-operators' shoulders or telling the secretaries which sides of the stamps to lick, can stand back and develop a policy that is more than a response to the needs of the moment. The brain structure that allows a bat to sense a network of twigs, fly through and snap an insect out of the air, does its work with superb efficiency; but having done that it is incapable of doing much more. It does not allow the bat to sit and reason out a way of dealing with an unfamiliar threat, such as human pollution of its food supply.

In this respect, it is interesting to compare the brain of *Homo sapiens* with that of another form of man (or near-man) the Neanderthaler. The Neanderthal skull was shaped differently from ours, sweeping back from heavy eyebrow ridges where ours rises in the near-vertical line of the forehead. This meant that it could not have accommodated the large frontal lobes that form a major part of our cerebral cortex. It also meant that Neanderthal man's skull was in that way similar to the skulls of the big cats. The streamlined shape of a leopard's skull is a very noticeable feature. It was certainly noticed by one East African hunter who shot a leopard at quite close quarters with a 0.275 Rigby rifle. As he walked towards it, the leopard jumped up and would have ripped him to pieces if his African assistant had not been covering the leopard with his gun and had quick enough responses to pull the trigger in time. The first bullet had deflected off the sloping line of the leopard's skull, leaving it momentarily stunned, but far from dead.

This streamlined shape is a common feature of all the top land carnivores. It is well suited to accommodate the parts of the brain used to control motor function, which are in the middle, and the regions dealing with sight, smell and hearing which are behind and below. For example, a sudden bright light triggers a response in the occipital region at the back of the brain and a message passes to the motor region, which in turn sends a message to the muscle of the iris to close down the aperture. If the light reveals a dangerous assailant, a whole range of additional messages is sent all round the interpretive and motor centres to elicit a physical response which may bring the bulk of the motor region into play. These two parts of the brain are used whether one is avoiding a tiger or a motor car, or catching a cricket ball or a mouse. They are fundamental parts common to all mammalian brains.

Neanderthal man might well have had those regions developed to a high degree of efficiency and it is significant that, judging from the fossil remains, he appears to have been carnivorous. Indeed, it has been suggested that a distinguishing feature between Neanderthal man and *Homo sapiens* is that the Neanderthals did not eat fish. Perhaps we should see him as a kind of 'action man', who migrated inland carrying with him his still large brain. He then competed with the top carnivores (who knows, he may have had eyes like a cat) but lacked the capacity for intelligence of his cousins with the high foreheads. He may well have become Robert Ardrey's 'killer ape' (Ardrey, 1961)! In the end he became extinct: probably when man did what men do so often – eliminated a rival.

Across the world man's close link with water and its foods is so obvious that its significance has been missed. The idea of a savannah origin is quite inconsistent with human physiology and in the next chapter we shall see further evidence for a human link with water more specifically related to his big brain.

9 The origin of the big brain

A problem with Hardy's theory concerns his timescales. If the main line of the evolving hominids did not go into the forests in the first place, Hardy's timing could be wrong. The more serious omission of both Hardy's and the classical view is that neither consider nutritional input in any qualitative way. No real mechanism is offered for the key to man's difference from other species: the big brain. The aquatic ape theory offers no better reason for the evolution of the big brain than does the savannah theory. It is all very well to say that all the savannah species lost brain capacity as their bodies became larger, but why did this law not apply to man? Why is there not the same evolution of large brained species on the savannah as there is in the sea? Could it be that our common love of the sea and its foods provides us with the clue? But if so, what does the clue tell us?

There is of course an old tale that fish is good for the brain, which may well date back to the time when monks kept fish ponds and were considered to be especially clever people. The notion of fish being good for the brain is, however, best captured by the humorist P. G. Wodehouse who wrote about Bertie Wooster, a rich but not very bright young man, and Jeeves, his manservant. Jeeves was totally brilliant and frequently had to rescue Bertie from indiscretions and other difficulties of his own making. On one occasion, a member of the Cabinet whom Aunt Agatha had wished to impress, was chased by a vicious swan and made to seek refuge on her boathouse roof. Bertie made a valiant effort to be the saviour but was also forced to flee to the same sanctuary. However, the approach of Jeeves stealing through the shrubbery was greeted with the utmost relief and confidence in the imminent rescue. After all, Bertie mused in order to reassure the Rt Hon. member: 'There are no limits to Jeeves' brain power. He virtually lives on fish.'

In the first 1.6 billion years of life on earth, the attention of the blue-green algae and bacteria was focused on DNA–RNA–protein chemistry with little regard for the lipids. However, about 600–500 million years ago there was a quantum leap into multicellular designs. This quantum leap in organisation utilised a new dimension in biochemistry: namely the fatty stuff called *structural lipid* which is used with protein to make the membranes on which this organisation depends.

Previously we discussed how the multicellular design meant specialisation which, ultimately, meant nervous control and finally a brain. Furthermore, the brain is critically dependent for its nourishment on a vascular system. Both systems are dependent on specialised structural lipid. This introduction of lipids as a new dimension in biological organisation for multicellular systems, suggests that they were as central to animal evolution as was oxygen: so it is hardly surprising to find today that both are used to the highest degree in the brain.

Whilst muscle accounts for the largest single mass in the body, it focuses mainly on protein. Hence whilst protein is important in muscle and minerals in bone, lipids were the new dimension brought in by evolution to serve nervous control and on which today both our brains and vascular systems depend.

The origin of our brain may well have something to do with the origin of its building materials: the structural lipids. Here we need to recapitulate the discussion which explained the superior night vision and nervous systems in the carnivores compared to the herbivores.

Different species build muscle protein with the same profile of amino-acids. This suggests the profile was established very early in animal evolution. By contrast the lipids used to build tissues of different species are different with one outstanding exception: the brain. The pattern of neural fatty acids is the same for mouse, squirrel, elephant and human. Importantly, the essential fatty acid component of the brain contains arachidonic and docosahexaenoic acids and not the parent essential fatty acids linoleic and alpha-linolenic and deprivation experiments have shown the requirement for them.

A paper entitled 'Nutritional influences in the evolution of the mammalian brain' (Crawford & Sinclair, 1972) argued that carnivores obtained significantly higher concentrations of these fatty acids in their diet which could explain the difference between the higher degree of development in the carnivores compared to the herbivores.

However, this data did not explain everything. The top land

carnivores have none the less lost an enormous amount of brain capacity. Compared to *Homo sapiens*, they have very small brains. The big difference comes in the jump from secondates to primates but none of them are especially carnivorous. The probable explanation may be that the primates grow slowly and have long gestation times allowing the foetus to be well fed by the placenta. However, as the top primates all have very small brains compared to ours, there is a need to explain that difference.

The classical view is that our brain capacity evolved in a straight line from a 450ml size (that of a chimpanzee) to the contemporary 1300 to 1600ml capacity on the basis of need. This view has had to wrestle with 'the fossil gap' which is otherwise described as 'the missing link'. However, brain growth is not concerned with straight line growth as in the growth of a leg bone. It involves volume increments where colossal numbers are implicated in the increments in sophistication. With a thousand billion cells in the human brain, each making thousands of connections with others, going from 450 to 1300ml or more is not a matter of straight line development: it is a huge, multidimensional jump.

To visualise the huge dimension across this stretch of development we can use the analogy with telephone lines and computers. A computer works on the basis of addressing memory to retrieve information for use. A 7-byte address bus using the numbers 0 and 1 allows you to locate 128 different addresses. The addressing range increases with the number of lines in the address bus, not as a linear increase but as a power function. A modern computer with a 32-bit address bus can instantly gain access to a phenomenal number of 4 gigabytes which bear no comparison to the straight line between 7 and 32.

A telephone exchange operates on a similar system and is something everybody knows about. We can use this analogy to compare three species: first the chimpanzee with 450ml capacity, secondly *Australopithecus* with a 700ml capacity and thirdly man with a 1300–1600ml capacity. A 4-digit number (the chimpanzee equivalent) will suffice as a means to locate all the people with telephones in a small village or town. At the time of writing a 7-digit number will reach all in a major city the size of London. However, you only need to go to an 11-digit number to dial the whole world. The linear increment from 4 to 7 to 11 seems small, but the volume it represents is enormous. The idea of going from the chimpanzee's brain volume to man's simply on the basis of need seems absurd. The direct line from the chimpanzees invokes too large a dimension. A line which did not lose its brains makes more sense.

In 1951 S. B. Leakey and Le Gross Clarke jointly published a monograph entitled 'The Miocene Hominidea of East Africa'. They showed that forms of primate ancestral to the gibbons of the Far East known as *Limnopithecus* had been present in East Africa during the early Miocene period. They were in fact contemporary with *Proconsul*. The gibbons had separated from the other ape stocks by that time and Le Gross Clarke and Leakey offered evidence that *Proconsul* possessed a number of characteristic features more consistent with a human than an ape. *Australopithecus* walked upright, which suggested an early separation of a hominid from the apes. It was this type of evidence which fuelled the idea of a direct line to modern man. No thought, however, has been given to the prodigious jump needed to develop even from *Australopithecus*, never mind from a chimp-sized brain to modern man.

Based on our initial evidence on carnivores and herbivores, Robert Ardrey argued that the big brain of *Homo sapiens* could be explained by the fact that *Homo sapiens* was a 'killer ape' whereas the chimpanzee was a vegetarian. However, the difference in development between the lion and the cow is not really big enough to explain the difference between the chimpanzee and *Homo sapiens* simply on the basis of eating meat. Something was missing and it was studies on the dolphins which supplied the clue.

Laboratory work showed that the brain did not just use any old neural fatty acid but a consistent balance of 1:1 of both the Omega-3 and Omega-6 families. So not only were these special fatty acids important to build brains, but the balance between the two families seemed relevant (*see* Table 3).

With the evolution of seed bearing plants introducing linoleic acid and its Omega-6 family of fatty acids it now happens that in general, land mammals have about three to six times more Omega-6 than Omega-3 fatty acids in their organs other than the brain. By contrast, in the marine environment, the Omega-3 fatty acids dominate. The codfish has a ratio of Omega-6 to Omega-3 fatty acids of 1:40 in its muscle membranes. So if this balance is critical to the brain, the mammals that had once lived on land but now live in the sea were almost certain to destroy the consistency and hence the argument.

Table 3

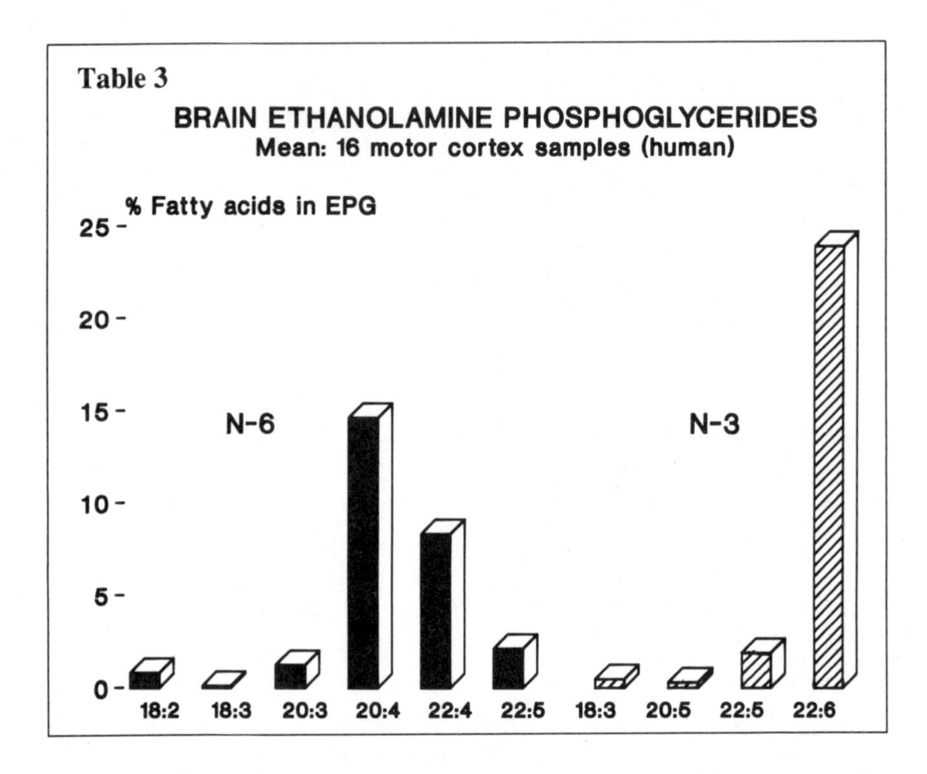

In collaboration with Dr Bill Perrins of the North American Fisheries and Glynne Williams from the Nuffield Laboratory, the dolphin was chosen for analysis because of its big brain and because it could be examined in captivity.

Studies on dolphins accidentally killed in tuna fishing brought to light the remarkable fact that their liver stores and muscle tissues did not have a 1:40 ratio like fish: they held a 1:1 ratio. The same was true for their brains. Biochemically, they were still land mammals living in a marine habitat.

Clearly, they had a special food selection technique and or metabolism, which enabled them to seek Omega-6 fatty acids in the marine food chain; the same fatty acids upon which they had depended as a land mammal. They had not become like fish. This result strengthened the idea that the balance was important for mammals and especially for their brains. Their liking for squid probably made a significant contribution as squid contain a certain amount of arachidonic acid.

Biochemically the dolphins had one important advantage. Unlike the land mammals, they had an abundance of Omega-3 fatty acids, particularly docosahexaenoic acid. Their disadvantage is the scarcity of the Omega-6 seed type, specifically arachidonic acid. One has to know where to look for them and the dolphin obviously knows.*

There is another absolutely fundamental difference between land and sea. The base of the food chain in the sea as algae and plankton starts with long chain Omega-3 acids (20 and 22 carbons) much closer to those used in the brain. On land, the start of the food chain for essential fatty acids is confined to short 18-carbon chain lengths. The difficulty of building the longer, more unsaturated fatty acids is apparent in the land food chain. Gorilla milk was virtually devoid of the long chain fatty acids. The marine food chain on the other hand is carnivorous and its fatty acids are virtually all of the long chain type preformed.

So perhaps here was a simple answer. The dolphin could have maintained brain capacity simply because the necessary Omega-3 fatty acids were available. So what about man?

Had man been a savannah species he would, like all other savannah species, have found it difficult to obtain the long chain fatty acids in his food, especially docosahexaenoic acid. He would have had difficulty in satisfying the nutrient requirements for his brain and visual system: which may well have happened to certain hominid offshoots.

Had he occupied the vacant niche between land and water things would have been different. He would have had available Omega-6 fatty acids from land seeds, protective anti-oxidants, and high

* A few words about the distribution of the neural fatty acids. Marine or Omega-3 lipids typically contain two important fatty acids: namely eicosapentaenoic (EPA) and docosahexaenoic acids. However, coastal resources like marine algae, shell fish, crabs, squid and the like also contain significant amounts of arachidonic acid, an Omega-6 fatty acid. Whilst Omega-3 fatty acids predominate in cold, and deep water species, the Omega-6 fatty acids can be found in warmer waters and in species living in estuaries and coastlines. Tissues of marine mammals are rich in docosahexaenoic acid but they also contain arachidonic acid. This they assumedly obtain from the coastal food chain. They would have been used as food by coastal hominids as they are for Eskimoes today. For example, seals often visit the land, to bask in the sun and have their pups which makes them an easy prey. A. J. Sinclair has shown, in warm waters the fish contain much more arachidonic acid than the cold water fish in the North Atlantic (e.g. cod and herring) which are closer to Bob Ackman's expert laboratory in Canada and, therefore, most frequently analysed. In addition to this, these coastal marine foods are particularly rich in trace elements and a wide range of vitamins.

concentrations of arachidonic acid preformed, from small land mammals, from freshwater foods and coastal seafoods as well as from marine mammals. He would also have had an abundance of the docosahexaenoic acid which is missing in the food chain of large land mammals. The land–water interface was an incredibly rich ecological niche which would otherwise have been unexploited. To achieve a ratio of 1:1 would have presented no problem.

Furthermore, nutrition is not just about fatty acids. The same location would have undoubtedly been a superior source of foods rich in vitamins and especially rich in trace elements. The interface quite simply provided the best of both worlds.

By contrast, in the savannah, evolution explored the greatest membrane aberration: the decrease of brain capacity in all its evolving inhabitants.

There are some contemporary streaks of evidence which add weight. First, there is the problem of vitamin A blindness, so prevalent in the predominantly land-based vegetarian populations of India. On the one hand, people will say that this is because they eat the wrong foods, and this is true. On the other hand, the high incidence of vitamin A blindness also shows that this choice of direction in food does not make it easy to meet the requirements for the visual system.

Secondly, there is the link between nutrition and the outcome of pregnancy. Perinatal mortality and handicaps involving improper development of the brain or nervous system are far more frequent in association with low birth weight or foetal growth retardation. Several studies indicate a link between perinatal mortality and low birth weight and nutrition of the mother.

The evidence from the follow-up study of all children born in a specific two week period in the UK from birth to 17 years of age, indicated that the risk of educational subnormality was many times greater in the lower socio-economic groups. It is unlikely that this large difference is the result of genetic differences between social classes.

Wendy Doyle has examined food intakes of pregnant mothers from two different socio-economic groups in London: Hackney in the East End and Hampstead in the rich North West. With smoking eliminated in her study, it was clear that those mothers who gave birth to babies of low birth weight were eating a poorer diet. Her data on food intakes in pregnant mothers from groups at high risk of low birth weight, small head circumference and associated handicap, imply that nutrition, not just during pregnancy but the habitual diet, is strongly correlated. It is not just what the mother eats whilst she is pregnant but her background nutrition which prepares her, possibly during and from puberty. So both Western and Third World evidence concur: nutrition may influence the outcome of pregnancy and hence the development of the brain of the next generation (Doyle, Crawford & Laurance, 1982).

There is new laboratory data describing a specific nutritional link with lipids and foetal growth retardation. Recent studies by Dr Gerard Hornstra in Maastricht, Holland, and at the Nuffield Laboratory, have described specific deficits of essential fatty acids in the umbilical artery (a piece of foetal tissue) at birth related to low birth weight, small head circumference and low placental size (Hornstra et al., 1989; Crawford et al., 1989).

Additionally, the English paediatrician Dr R. W. Smithells has studied neural tube defects which are far more frequent in the lower socio-economic groups. If a mother has a spina bifida or anencephalic baby, there is a strong chance she will have another. Dr Smithells divided mothers who had given birth to babies with neural tube defects into two groups, and supplemented their diets with either a dummy pill or a vitamin supplement pill. When they came to have their next baby, the incidence of neural tube defects was seven times less in those who had had the vitamin pill. These and other studies indicate that, just as with the animal experiments, maternal nutrition during and even before pregnancy is a determinant of foetal growth and brain development (Smithells, Shepherd & Schorah, 1980). This stretch of evidence from the evolution of the brain to present day experience, tells us of the singular importance of the women in our society and our future.

Thirdly, there is evidence specifically related to inland nutrition retarding mental capability. Millions of years of snow and rain had depleted the soil of the Alps of its elements, and, in particular, created an iodine deficiency which caused goitre and cretinism. Such a problem could not have arisen in people living on seafoods. Oysters, mussels, abalone, scallops, crabs, lobsters and, indeed, all the coastal and estuarine foods are outstandingly rich in trace elements.*

* The French insistence on drinking mineral water is interesting in this respect. The water from the famous spring at Evian takes 17 years to work its way through the mountain rocks. One suspects it must pick up some trace elements en route! It is also interesting that the incidence of heart disease is higher in soft water locations.

Everyone accepts that nutrition can retard body growth or, indeed, be responsible for promoting it. Is there any reason why the brain should be totally immune to this same force? Cow's milk is known to be rich in nutrients important for body growth. We can accept that the Japanese started to become taller with the introduction of cow's milk and meat to their diet. Is this in any way different from conceiving foods which sponsored brain growth?

None of this contemporary evidence proves nor disproves the idea of an aquatic origin for *Homo sapiens*. However, the experimental evidence, the socio-economic and cross-cultural data in humans all indicate that just as the body can be manipulated, nutrition can also manipulate the brain. The comparative evidence of difference in availability of the same neural nutrients known to affect brain development in experiments, and the huge contrasts in brain size, speak of long range forces. This comparative evidence, with the loss of Omega-3 fatty acids from the land food chain, makes it exceedingly difficult to comprehend a savannah origin and offers *Homo aquaticus* as the line which led to modern man.

It should be remembered that animal life first originated in the sea where there was an abundance of Omega-3 fatty acids. We now find that these fatty acids are key components of the photo receptors in our eyes and synaptic junctions in our brains. However, the animals associated with the first phases in evolution did not build big brains, they were egg laying. Seed-based fatty acids were yet to come in a concentrated packet. At the end of the Cretaceous period, the dinosaurs collapsed and the soft, flowering and seed-bearing plants emerged and brought with them linoleic acid and a new group of animals, the placental mammals, with their requirement for the Omega-6 family of essential fatty acids.

This new nutrient dimension did two things: it made a 1:1 ratio of the two families of fatty acids easier to attain; simultaneously it became essential for the special reproductive and vascular process of mammals which, we suggest, ultimately made possible the evolution of large mammalian brains.

In summary, we can now suggest a chemical mechanism for the origin of the human brain. The evolution of seeds, and then the evolution of placental mammals, complimented what was already present in the sea, and opened the potential. As the land species developed bigger bodies, so they outstripped their ability to build the brain's necessary molecules and their relative brain capacities shrank. Man did not grow a big brain, he simply retained the relative brain size he had as a small mammal. All he had to do was to find an ecological niche which provided an abundance of both the neural fatty acids and other nutrients to maintain his brain capacity as he grew bigger. The chemical evidence tells us that this ecological niche was the land–water interface occupied by *Homo aquaticus*.

The ecological evidence also suggests that it would be strange if the land–water interface had not been occupied by evolving man. With the migration of the marine mammals to the sea having finished over 5 million years ago, this incredibly rich food resource would otherwise have been left empty. Why would nature exploit the poorest desert fringes in the form of the desert rat or the high mountains with the snow leopard and leave such a rich food resource untapped?

With the brain capacity shrinking in all land mammals, it would be trite to say that it was eating fish and seafoods which filled the gap and enabled *Homo aquaticus* to retain his brains instead of losing them. We suggest that human evolution, like that of previous epochs, was substrate-driven. By using the land–water interface, *Homo aquaticus* need have been neither wholly aquatic nor wholly land based: he simply enjoyed the best of both.

And so we can now resolve the final puzzle, the missing link. The missing link is, in a sense, not missing. The missing link is fossilised in human history, the origins of civilisations, behaviour, food and geography, love of the sea and water, the universal attachment to sea and river foods, and it is all around us: it is so commonplace that it is overlooked. It is the fossilised evidence for which we have been searching and it is everywhere to see.

10 Food and civilisations

The interaction between nutrition and genetics had played its part in shaping the mammals. Man emerged as a product of the last great evolutionary experiment of nature to date. Let us now continue the discussion on the role of food in evolution by examining the way it worked on different civilisations.

However one looks at history, the foods man ate throughout 99.8 per cent of his history and those which he eats today are different in many ways. There is only one exception and that is seafoods. Our relationship with marine and freshwater foods is still that of a hunter-gatherer. In this respect, we have made little advance since the days of the Pleistocene. This argument makes one wonder if people who still rely on fish and seafoods have retained any advantage? In so far as the commonest degenerative disease of Western populations, namely heart disease, is concerned, the answer to that question would be yes.

If food was important in evolution, the recent deviations from our 'baseline' foods would be expected to throw up some interesting developments of a beneficial or, perhaps, unwelcome nature. We have seen how evolution can go backwards as well as forwards. It can be predicted that the different changes in different parts of the world will have produced different effects. If the concept of substrate-driven change is valid there will be evidence in history and today of food changing man in shape and performance. It is this proposition that we will now examine in the remaining chapters.

The popular view is that man evolved as a hunter and gatherer. This notion is not inconsistent with the idea of *Homo aquaticus*. As man became physically larger and able to move long distances, subgroups would undoubtedly have explored inland regions, probably making use of the waterways. The motivations would have been similar to those which made people set off on the transoceanic migrations, the feasibility of which was demonstrated by Thor Heyerdahl. Indeed recent genetic evidence implies that modern man can be traced back to a single woman who lived 200,000 years ago, and the appearance of intelligent cultures on the waterways would suggest water, rather than land, migration.

Over 50 of the land-based hunting and gathering societies came into the 20th century but no more than a handful still survive.

Around 80 scientific papers, articles and books have been written on these people and reviewed in an article on 'Paleolithic Nutrition' by S. Boyd Eaton and Melvyn Konner in 1985. Their work is based on the assumption that the land-based hunting and gathering communities really represent the pattern of human existence throughout its evolution. Yet if it did, why did so few remain to this century and even fewer survive the first 50 years of it? Maybe land-locked meat-eating and gathering was a dead end. They certainly did not prove to be the 'fittest'.

If one asks that conventional question about the 'fittest' today (one could use the quantitative measurement of the financial markets, exchange rates, trade surpluses, drive and financial intelligence), the Japanese recommend themselves as being pretty fit. Based on the growing budget deficits of 1981–9 in the USA and the growing surplus cash in Japan, the USA has collapsed into a position of such debt that the Japanese could just about buy up the USA in the remaining 11 years of this century. The Japanese can be said to be the one remaining successful hunter and gatherer culture, but they specifically hunt and gather the sea. They eat five times more fish than the British. To replace their seafoods by agriculture, they would require seven times their present amount of land. No nation would have indulged itself to that extent had it not been led on by a superabundance. Is it possible that nutrition has played a role in shaping the success of nations?

Let us start again from the beginning. Sanford Miller and Marylin Stephenson from the Centre for Food Safety and Applied Nutrition of the Food and Drug Administration of the USA opened an address to the European Group of Nutritionists in Evian in March 1986 with the following words: 'Food has always been a determinant of civilisation. It is a truism that only those cultures which have been capable of controlling their food environment have been able to survive.'

Civilisations, like epochs of evolution, passed through the same five phases, and, like species, they were stimulated, or checked, by food, energy and raw materials. As in evolution, two great influences can be seen at work: the driving force of nutrition followed later by the struggle for resources which favours some and, in the end, destroys those that can no longer find their necessities in the environment they have helped to shape.

In the beginning, in the warm regions round the coasts and estuaries, those developing hominids, destined to become man, would have been close enough to the equator not to be bothered by extreme variations of climate. Their physiology had evolved over a period of between 5 and 50 million years, tuned into an ample water supply and a rich, diverse diet. This diet coupled with plenty of exercise would have provided the basis for fitness, good physique and intelligence. A rich life around estuaries with little competition, except for the transient migrations of species which ultimately became the marine animals.

Eventually such prosperity would have led to an increase in physical size and an increase in population. The best niches would have filled up, and just as the Polynesians migrated in recorded history, early splinter groups would have spread out into new areas, some with less hospitable climates, further north and further inland. Some would have had to learn how to live in cold conditions, using meat and offal from the larger marine and land mammals, cutting and sewing skins to protect their hairless bodies. As they moved north they would have been more and more affected by the long cyclical changes of climate that produced the ice ages and the warm interglacial periods between them. Close to the edge of an advancing or retreating ice sheet, life is dominated by its movement. The tides of human population that flowed around the globe were largely a response to that impulse.

It is now thought that there have been seventeen glacial and interglacial periods in the last 1,700,000 years, about one every hundred thousand years (Sutcliffe, 1985; J. & K. P. Imrie, 1979). Their causes are complex and not fully understood: one of the simple reasons is the tilt of the planet on its axis of a few degrees roughly every 40,000 years. There are many other causes connected with variations in the amount of heat the earth receives from the sun and the amount that can escape (e.g. the greenhouse effect created by increasing carbon-dioxide). Ice obviously reflects the sun's rays but also represents much of the free water of the planet, frozen solid and having the effect of drying out the planet, increasing the risk of desert formation.

Darlington relates how the last great glacial ice age froze so much water into the icecaps that the level of the sea fell by over 600 feet. With much of the present sea floor exposed, islands were joined into continents and land bridges linked Asia both to Australia and America. Across the bridges came migrating bands of primitive humans, those from the north, probably driven by the intense cold to look for warmer, more fertile regions. They would have followed the movements of plants, land animals and fish, whose habitats had shifted south. Mongolian peoples, bison, mammoth and antelope moved across the Bering Strait bridge, passed through Alaska, and over the next 8000 years spread all over the Americas, although it was only some 2000 years ago that the Eskimoes settled in Alaska and not until AD 1300 that they reached Greenland.

As the glacial passed and northern climates grew warmer, the retreating ice gave way over a large band of the earth, first to great grassy plains and then to forests. In the more temperate climate prevailing, the stage was set for civilisations to follow. When they did so it was on the basis of a new way of life – settled agriculture – and what first impelled men in that direction may well have been the onset of a world-wide disaster. The story runs something like this.

When early man spread northwards from the equatorial regions, the exposure to extremes of climate is likely to have forced him to develop a more omnivorous approach. Unable to rely any longer on an abundance of suitable foods it is supposed that he needed to adapt his demands to the supplies available. Eaton and Konner suggest that, in particular, the high-energy food that exposure to cold demands is most readily found in meat perhaps implying that man depended on animals for more than half of his food for over two million years (Eaton & Konner, 1985). At the beginning this meant eating raw flesh, since fire was not then widely used.

Our laboratory data on wild animals show that 100g of reindeer meat contains only 107 kilocalories. Wild meat is characteristically lean and the carcass contains little fat. By contrast, 100g of average Aberdeen Angus beef contains 637 kilocalories and average fat pork 670 kilocalories. It is only as a result of modern, intensive, high-energy feeding systems that mountains of excess fat are produced on our domestic animals; the trouble is that people often make the mistake of assuming that this is the rule: it is not. Hence it is wrong to assume that man would eat meat from land based animals to obtain a high fat, high calorie diet. Whilst it is likely that primitive man killed and ate venison, moose and reindeer, the food resource which would have provided a high fat, high calorie diet would have been the cold-water marine species, which adds more evidence for the aquatic link.

Some people, like the early Vikings, learned to exploit both marine and land habitats. Other groups like the Japanese and the Eskimoes remain specialists in seafoods to this day. Eskimoes pray to the spirit of the seal they have just killed in gratitude for sacrificing its life to provide food, blubber and skins. The attitude of people who had come to dwell inland was less caring. Until recent times, North American Indians drove herds of bison to their death over cliffs and Laplanders did the same with reindeer. Far more animals must have died than the small populations could have used but a balance must have held for a long period. As populations grew and weapons became more sophisticated that balance was destroyed. There came a time when a change in climate started to alter the flora, so possibly the plants on which the food animals lived began to die out. At about this time there came hunters armed with the new 'flute-pointed' spears. Whether it was climatic change, superior weapons and hunting techniques or demographic change, or a combination of the three, the effect was what became known in the US as the Pleistocene Overkill (Martin, 1966).

As human populations increased, man exploited inland regions and land plants and animals came under pressure. Early man then learnt to domesticate plants and animals. This change in the method of getting food was of the greatest importance. C. D. Darlington in his *Evolution of Man and Society* relates the origins of some of our domestic plants and animals (Darlington, 1969). The 'Neolithic Revolution' that produced agriculture was the basis of civilisations that followed, and it seems to have occurred at about the same time, around 10,000 years ago, in widely separated places in Western Asia, in America and elsewhere. In isolated areas across the world, separated from one another by mountains, forests and deserts, the beginnings of settled cultivation started to appear.

The soil was virgin and the climate now conducive to this form of exploitation. In North America the first men to arrive found vast herds of herbivorous mammals, including camels, horses, mammoths and bison. Within 12,000 years 35 genera were exterminated and the great mammals vanished, all but the bison, caribou and deer. This rapid series of extinctions compares with only 13 genera of large mammals lost in the preceding one or two million years. In Africa, within the last 100,000 years at least 26 genera of large mammals have disappeared. The crash seems to have taken place world-wide and it forced two consequent changes on mankind, a change of diet and a change in the method of getting food.

Mutated and hybridised strains of cereals, beans, squashes and various other crops were selected to produce the cultivated forms. Different peoples made use of different plants, and in the same way the first animals to be domesticated differed from place to place. In the old world, they are likely to have been sheep and goats, followed by cattle and pigs, with the horse coming later. The use of domesticated animals spread from Western Asia into the fertile crescent round the eastern Mediterranean and eventually across Europe as far as Britain. The first agriculturalists probably came to Britain from Gaul about 4000 years ago, bringing with them the small, long-haired neolithic ox, a small number of sheep, and later the pig and horse. In South America the Incas domesticated the wool producing alpaca, and the llama as a beast of burden. Throughout America there was a shortage of domestic animals due to their over-hasty extinction by hunters between 12,000 and 8000 years ago (Mangelsdorf, 1964).

The evidence suggests that the spread of crops was not due to regular traffic on land but rather took place by sea, with specimens carried by fisher-folk and accidental castaways. In this way the pineapple is thought to have crossed the Caribbean from Brazil to Mexico. There may also have been transoceanic movements on a larger scale. Thor Heyerdahl has argued that the sweet potato, carried as food by navigators, assisted them to cross the Pacific to the Polynesian islands, to transplant their crop and to create a megalithic society first on the Marquesas and later on Easter Island (Nishyama *et al*, 1962).

In ways such as this the basis of agriculture was laid and its practices spread over the world. As the area under cultivation grew it steadily pressed back the territory where the remaining wild species of animals and plants were available to those who continued in the old practices of hunting and gathering. The age-long conflict between the farmer and the primitive tribesman had begun. Few societies or opportunities now remain today to represent the old way of life.

The more people took up the practice of agriculture, the more dependent on it they became, and the more vulnerable to any disruption. This vulnerability was made more precipitous by the fact that, under cultivation, the number of plants and animals used as food greatly decreased. As hunter and gatherer man used thousands of plants and animals but, as Professor Arnold Bender of Queen Elizabeth College, London, has pointed out, of all the vast numbers of wild plants that cover the earth 'only 3000 species . . . have ever been used for food; only 150 of these have ever been commercially cultivated, and the world lives on about 20 main crops' (Bender, 1980).

Just as agriculture solved the problems of the Pleistocene Overkill at the cost of making people permanently dependent on it, so in turn it made civilisation possible and the agricultural industry indispensable. As urban centres and their agricultural lands grew, so space became a limiting factor. The boundaries between communities would have become important and authorities arose to defend them and to impose order within them. The foundation of hierarchical societies was being laid. Nations, religions, warrior orders and social classes grew up.

In all this history, it needs to be remembered that up to the invention of agriculture, the species that was destined to be wildly successful, perform brilliantly (in evolutionary terms) and to dominate all living things, lived on food that was wild. That is, man has relied on wild foods for *99.8 per cent* of his time: even more, as at first wild foods played an important role alongside cultivation. Agriculture opened up a new mini-epoch, as progressive in terms of increased productivity as the invention of the boat. It worked well in certain climates, but disastrously in others. As far as man's physiology is concerned it is a new experience.

The earliest known civilisations arose in the river valleys and estuaries. The land was colonised by agriculture practised with the help of water, silt, and food itself from rivers and lakes left behind by the retreating ice, in areas originally well provided with tree cover.* Inhabitants of the Indus Valley were advanced agriculturalists by around 3000 BC, using irrigation; excellent potters using the wheel; and skilled bronze casters. Another region was in western Asia, where recent archaeological discoveries show that farming villages existed by 7000 BC, spread over a wide arc spanning the western flanks of the Zagros mountains in Iran, through Iraq and southern Anatolia and south into Palestine. Large amounts of plant remains have been found in several regions and it is now well established that neolithic agriculture in western Asia depended on three species of cereal, einkorn wheat, emmer wheat and two-row barley, which are thought to be the ancestors of our own cereals. What was probably the earliest civilisation of all arose in that region, in Sumer in the low land between the rivers Tigris and Euphrates. The annual flooding of the rivers brought not only water but silt, rich in minerals and other essential nutrients of the soil, and that water was spread through complex irrigation networks controlled by elaborate bureaucracies. It also fertilised the estuary and the marine food chain. Political systems evolved, new occupations were created and there were increasing divisions of class. Representational art was added to the earlier decorative forms, and the significance of public buildings was expressed in elaborate and imposing architecture. Armies drove into what is now northern Syria and Anatolia where there were sources of lead, silver and gold. They felled timber in the Amanus Mountains and floated the logs down the Euphrates to build their palaces.

It was too good to last. Already, in pre-biblical times, Egypt was being hit by recurring famines typical of Phase 4. An inscription in a tomb near Luxor describes a famine around 2180–2130 BC:

> All Upper Egypt was dying of hunger, to such a degree that everyone had come to eating his children, but I managed that no one died of hunger in this home. I made a loan of grain to Upper Egypt: I kept alive the House of Elephantine during those years.

A century of so later the 'Prophecy of Neffurty' includes this passage:

> The river of Egypt is empty, men cross over the water by foot. Men search for water on which ships may sail. The south wind drives away the north wind, and the sky has only one wind. The birds no longer hatch their eggs in the swamps of the delta. Foes are in the east; Asiatics are come down into Egypt.

Similarly, a man called Hekanakht wrote to his family on a business journey:

> The whole land is perished, but you have not been hungry . . . When I came hither southwards I fixed your rations properly . . . our food is fixed for us in proportion to the flood. So be patient (Lamb, 1971).

There can be little doubt that the Egyptians colonised the Nile valley because of its fertility. (They would not have chosen to colonise a desert!) Into the pattern of impending doom, stories of famines were commonplace by biblical and pre-biblical times. Could the advent of agriculture all along the Middle Eastern band which comprised Sumer, Palestine, Syria, plus Egypt, have combined to precipitate periods of drought and famine? Was this the first example of over-exploitation of land resources leading to climatic change and nutritional collapse?

Some consider that agriculture itself may have been at least partly responsible for climatic change. Joseph Otterman of Tel Aviv University, has suggested that, as domestic animals trampled and overgrazed much of the Middle East they may have encouraged a growth of deserts over a large part of what was once the fertile crescent (Otterman, 1977). A critical part was probably played by the destruction of tree cover.

It is commonly said that grass lies at the base of the food chain and in temperate climates which get rainfall all year that largely holds true. But arid, semi-arid and dry tropical climates constitute about one third of the world's land mass. They get their water – when they get it at all – in vast monsoon deluges. In such countries, the trees of forest, woodland and bush play an essential role in controlling the supply of water, acting like a great sponge when the rains come, holding the water, drawing it up, and letting it out in water vapour which helps many of the small plants living under their protection as well as going to form cloud which affects not only the aforested area but areas around it. Trees also, through their deep roots, allow surplus water to drain to the deep water tables instead of rushing around wildly on the surface, taking the soil and its trace elements with it. In such countries trees are the base of the food chain, for many animals, as well as people, live on their products and the plants they shelter. Trees are a crucial link between rain cloud, river, humidity and soil. Grass has shallow roots and grows but a few inches tall. By contrast trees fill a very much larger three-dimensional space.

Darlington examines the rise and fall of successive civilisations. He draws a direct link between the replacement of hardwoods by olive trees and the success and failure of the successive Mediterranean civilisations. The five-phase cycle can be seen very clearly in the history of the Minoan civilisation on Crete, which lasted for about a thousand years from 2500 to 1500 BC. It was one of the most impressive civilisations the world has seen, with great palaces and citadels and a navy holding sway over the entire Mediterranean. Darlington describes how its story begins with people coming to an island clothed in hardwoods and with a rich potential for agricultural fertility. Being great sailors and ship-builders, they cut down the trees to build boats, as well as for fuel, houses and palaces, and they cleared much of the land and planted it with olives.

Now the olive tree is no substitute for the old forest trees in its effect on the environment. The big, deep-rooted hardwoods like oak bring up moisture from the deep water tables far underground and slowly transpire it through their leaves into the atmosphere. Along with the water come minerals and trace elements from the deep soils and rocks, which are incorporated into the topsoil when the leaves fall and rot down into humus. In the shade of the trees grow herbs, sedges and other plants that can take moisture from the air. The olive, by contrast, has shallow surface roots which draw the water and nutrients out of the topsoil. It has hard glossy leaves that retain moisture and reflect back the sunlight. Its effect is to impoverish the soil, to discourage ground-cover plants and to leave the earth exposed so that when the rain comes it is easily washed away. Along with the goat, the olive tree must bear a large part of the responsibility for turning the Mediterranean lands from a richly forested, fertile paradise into the dusty countryside of thorn-scrub and marginal agriculture so many of them have now become.

With the depletion of the soil's fertility come poor nutrition and weakness. In Darlington's words:

> Crete may even in its last stages have been short of native timber. In these circumstances, when foreign invaders came, they found a weakened nation living on its past, full of treasures but now vulnerable to the invader. When they came there was no recovery. The princes departed and pirates took their place. The natives, the lower orders, the Eteo-Cretans, ultimately came into their own to wait for the next conquerors. The impoverished centre of a peaceful world had no place in the warlike times that lay ahead. (Darlington, 1969)

The final destruction of Minoan culture seems to have been brought about by an earthquake associated with a tremendous volcanic eruption in which the island of Santorin, 70 miles off the coast of Crete, almost completely disappeared. Some people have speculated that it was from this eruption that the legend arose of the lost continent of Atlantis, a great civilisation suddenly overwhelmed by the sea. But however destructive the earthquake may have been, it surely does not fully account for the fall of Minoan Crete. Other peoples have suffered from violent earthquakes and recovered; they have been invaded and have repelled the invaders. Darlington considers that the ultimate cause must have been the weakness of the Cretans themselves: the cause lay in their destruction of their own environment, in their growing dependence on imports and in the vulnerability that involved. If they had restricted their demand on their land, conserved a good proportion of its forests, maintained the fertility of its soil and somehow held their own population in check, who knows how long their civilisation might have endured? As it was, the great Minoan empire dwindled into obscurity. In 1450 BC Knossos was sacked by Myceneans from the Greek mainland and it was to that mainland that the centre of power now moved.

In Greece, as in Crete, the hills were generally wooded, and the plough followed the axe. The grandeur of the forests gave way to a system of agriculture similar to that of most Mediterranean countries; wheats, barleys and millets were grown, and olives, vines, figs and various other fruits and vegetables cultivated. The livestock included most of the animals we have today.

Some of the first evidence of climatic change, comes from Aristotle in Greece:

> The same parts of the earth are not always moist or dry, but they change according as rivers come into existence or dry up. And so the relation of land to sea changes too, and a place does not remain land or sea throughout all time . . . But we must suppose these changes to follow some order and cycle.

Around 800 BC the population had grown enough for a shortage of land to be felt. The Greeks sent out scouting parties to look for new places to settle and where they came across them they founded colonies. In the search for new sources of raw materials their trade greatly expanded and their culture entered the apparently buoyant and flourishing period of Phase 3. At the height of this phase Plato spoke out in Athens, warning its people of the consequences of soil erosion and urging them restore the broad-leaved woodlands. In Section 3 of the *Critias* he discusses the history of Athens and its surrounding land before his own time and describes its agriculture and manufactures. He draws attention to the relationship which had existed between hardwood trees and fertility, comparing the situation then with what he could see with his own eyes:

> But in those days the damage had not taken place, the hills had high crests, the rocky plains of Phelleus were covered with rich soil and the mountains were covered by thick woods, of which there are some traces today. For some mountains which today will only support bees produced not so long ago trees that when cut provided roof beams for huge buildings whose roofs are still standing.

At that time, he believed, the soil was more fertile than that of any other country.

> As evidence of this fertility we can point to the fact that the remnant of it still left is a match for any soil in the world for the variety of its harvests and pasture. And in those days quantity matched quality.

He recognises the use of trees as food for animals:

> And there were a lot of tall cultivated trees which bore unlimited quantities of fodder for beasts.

He is even remarkably penetrating in seeing the relationship between water and deep-rooted trees:

> The soil benefited from an annual rainfall which did not run to waste off the bare earth as it does today, but was absorbed in large quantities and stored in retentive layers of clay.

He sums up:

> This, then, was the general nature of the country, and it was cultivated with the skill you could expect from a class of genuine full-time agriculturalists, with good natural talents, an abundant water supply and a well-balanced climate.

His description of the scene in his own day is equally revealing:

> The Acropolis was different from what it is now. Today it is quite bare of soil which was all washed away in one appalling night of flood . . . The soil washed away from the highland in these periodical catastrophes forms no deposit of consequence as in other places, but is carried out and lost in the deeps . . . You are left (as with little islands) with something rather like the skeleton of a body wasted by disease; the rich, soft soil has all run away leaving the land nothing but skin and bone.

The replacement of the fertility-giving hardwoods by the olive and by shallow agriculture had again done its work. If Plato were to be brought back today he would weep: after two and a half thousand years we have not yet learnt the simple message. His words are a poignant reminder of man's greatest weakness: an arrogance in the face of nature.

The history of the next great Mediterranean civilisation, the republic and empire of Rome, is another classic instance of the five evolutionary phases. Rome started as a nation of farmers (a period to which many later writers looked back as a golden age). By about 270 BC she had become the foremost power in Italy and her growth gave her both the ambition and the need for overseas influence. This brought her into conflict with the other people who were making a bid for dominance of the Mediterranean, the Phoenicians. From their home in what is now Lebanon, with its cities of Tyre and Sidon, the Phoenicians had colonised much of the north coast of Africa, where they had vast estates, their greatest colony being Carthage. They had been competitors of the Greeks in their colonising days, as they now were of Rome, and their power stretched from the Straits of Gibraltar to Sicily, at its height the granary of the Western Mediterranean. The Lebanese homeland had been conquered by Alexander and the lands remaining to them had become treeless and over-grazed by livestock which included one distinctive newcomer, the camel. The Punic Wars (264–202 BC) were the struggle for supremacy between the declining power of Carthage and the expanding power of Rome as it entered its Phase 3. Hannibal's march over the Alps was an attempt to hold back Roman encroachment by conquering it at source and his defeat marked the end of the power of Carthage.

The olive, which together with the vine had been introduced into southern Italy by the Greeks around 700 BC, worked its way up to Rome by about 150 BC and seems to have been helped by a change in climate. The Roman writer Columella (AD 30–60) noted that:

> . . . the position of the heavens has changed . . . regions which previously on account of the regular severity of the weather could give no protection to the vine and olive stock planted there, now that the former cold has abated and the weather is warmer, produce olive crops and vintages in the greatest abundance (Lamb, 1971).

By the time Rome reached the height of her empire, she had become heavily dependent on imported foods. Grain was brought in from places as far away as Spain and Britain, but above all from North Africa. At the peak of the Empire, half a million tons of grain were being imported each year from the granaries of Rome in North Africa. But the farmers of North Africa knew little about soil husbandry and the same principle as described by Plato and by Darlington operated here. The trees had been cleared, and like the dust bowls of North America, the topsoil and its fertile mineral supply was being blown and washed away. Harvest failures became frequent.

Rome itself became full of unemployed people living on a dole of bread, olive oil and wine, and it was no longer possible for most ordinary people to make a living off the land in Italy. More and more small landowners, including war veterans who had been given parcels of land, found it impossible to pay their taxes and still keep a subsistence for themselves. Increasingly, they sold out to large landowners until eventually most of the land was owned by a few hundred families, who found it more profitable to put it into pasture and raise cattle.

In her long protracted Phase 4, Rome struggled to hold the frontiers of her unwieldy empire but, one by one, lost control of her sources of supply. Her barbarian neighbours were in their own expansionist Phase 3 and the soft, decadent society of Rome could not hold out against time for ever. She was hindered by the increasing difficulty of getting crops off the great North African wheatfields as their topsoils degenerated and harvests failed. The elaborate aqueducts and storage tanks she built were no real substitute for a sound ecology. Eventually Rome, having outstripped her resources, was severed from her food supplies by a simple blockade of the Tiber and fell in the undignified time of a week.

There are those more expert than we who have discussed this history and that of other civilisations and we shall remain content to offer these few examples from the two authorities at the extremes of written history: Plato and Darlington. We leave the reader to fill in the details as to how the ecology and the nutrition it produced would have played its role. Similar patterns are easily recognised today, portraying how history repeats itself.

A hopeful sign of the conservation scene is that conservation is no longer a concern of fringe groups. The International Union for the Conservation of Nature and Natural Resources, in co-operation with the United Nations Environment Programme, the World Wildlife Fund, the Food and Agriculture Organisation of the UN and UNESCO, produced in 1980 a 'World Conservation Strategy' which is aimed at achieving three main objectives: (i) to maintain essential ecological processes and life support systems, such as soil regeneration and protection, the recycling of nutrients, and the cleansing of waters on which human survival depends; (ii) to preserve the genetic diversity (the range of genetic materials found in the world's organisms); and (iii) to ensure the sustainable utilisation of species and ecosystems, notably fish and other wildlife, forest and grazing lands which support millions of rural communities as well as major industries. The project is backed by 450 governmental organisations and hundreds of scientists.

Just as oxygen was necessary for the evolution of animals, so the right climate and soil was necessary for agriculture, upon which civilisation now depends. In ten or twelve thousand years farming has spread across the globe exploiting one location after another. In 1492 Columbus rediscovered America. In the 500 years that followed, the immigrant population has sprung from zero to 200 million and is now dependent on large-scale agriculture. The speed of that deforestation and agricultural development was phenomenal. The virgin soil responded with rich harvests. Much of the world now depends on the American economy and its produce. But will that also prove too good to last?

Today there are no more continents to exploit: ironically *Homo aquaticus* may well have to return to his roots in the sea where our method of exploitation is still the primitive hunting and gathering. At the beginning of agriculture no one had any real idea as to the effect deforestation and agriculture itself might have on the planet's climate. But the evidence was identified by Plato 2500 years ago. He and Darlington understood the power of this effect on soil, nutrition and civilisation.

Today, the topic of reckless deforestation, acid rain, the greenhouse effect and destruction of the ozone layer, is recognised as a concern affecting all who live on this planet.

11 Departures from a common baseline

The test of nutrition or substrate-driven change is that a biological expression of change would follow a significant alteration in food structure. This is exactly what happened.

Once the breakthrough to agriculture and civilisations had been made, different groups of people developed into different civilisations. As their individual cultures evolved so came different developments in foods. We can learn for example from residual landlocked hunters and gatherers on the one hand, and the Eskimoes on the other, that neither have the Third World health problems such as kwashiorkor and vitamin A blindness, or those of the Western countries such as heart disease, lung and breast cancer. It is apparent that Americans are taller than Chinese and Japanese, and that Americans have heart disease whereas the Japanese do not. The simplified conclusion is this:

> *Starting from a common baseline of hunting and gathering in sea and on land, different cultures developed in different directions and now have different problems. In one place people developed rice and in another fat animals.*

Up to this very late time in human evolution man had lived on *wild foods* from land, rivers, lakes and sea. Human physiology for millions of years was adapted by and to those foods. With the exception of seafoods, fish and game, the process of the domestication of plants and animals started no more than 12,000 years ago and has led to across the board changes in chemistry, nutrient concentrations, balance and availability. At the beginning, the changes were slight. Wild foods continued to play a large role in the diet and pleasure of living. As wild resources have diminished in recent centuries, the aristocracy made certain that those pleasures were unequally shared.*

* In England, a forest was defined by Manwood, the early authority on the subject as: 'a

In England the introduction of enclosures for animals in the 17th century, genetic manipulation and the application of the tools of the industrial revolution to food processing led to sudden and dramatic changes in food composition. Have these had any effect on man's own development? Man, it is often claimed, 'has adapted best and will therefore continue to adapt'. But if so, the theory we are presenting will ask, in which direction will he adapt and how long will it take?

First, let us look at the direction taken in India. The development of and dependence on rice led to the exclusion of other foods (and hence nutrients) and consequent related disease. The exclusion of green foods, liver and other sources rich in beta-carotene and vitamin A resulted in blindness. Today much of the world's nutritional blindness is in India. Beriberi and pellagra resulted from the practice of removing the nutrient-rich germ and coating from rice. The specialisation of rice had brought with it a serious penalty. Here is evidence that change in food composition matters.

Sir Robert McCarrison was one of the pioneers of nutrition research to recognise the link between food and disease. As a doctor, he went to work amongst the Hunza in the Himalayas where he was dismayed, astonished and then uplifted because there was little work to do. He discovered instead the importance of nutrition to health, a subject to which he was to devote the rest of his life. He was stunned by the comparison of these outstandingly healthy people with those living on polished rice, and indeed his rickety fellow countrymen at home in the UK. He concluded that the 'unsophisticated foods of nature' provided for good health and by contrast many of the modern foods were the cause of the opposite. McCarrison's unsophisticated foods of nature are the equivalent of the wild foods which provided nourishment throughout human evolution.

To test his conclusion, he, like Eijkman, compared the effect of feeding experimental animals on the diets of the Hunza or on the

---

circuit of woodes . . . and privileged for the abiding of wild beasts and fowls . . .'. For centuries, the one passion of the English landed gentry seems to have been the pursuit of game. As human populations grew so the demand for wildlife grew and the artistocracy introduced repressive Forest and Game Laws to keep the pleasure and food to themselves: 'no punishment was too terrible for the poverty stricken wretch who poached to supply his needs' (*Everyman Encyclopaedia*, vol. 6, p. 384, 1913). Even as late as 1827 an Act was passed punishing the killing or wounding of deer with seven years' transportation.

diets of other indigenous groups where disease was rampant. During one period of two and a half years (1925–7) McCarrison kept 2463 rats in conditions of perfect hygiene but fed on faulty diets which emulated faulty human diets, and an average of 865 control rats fed on his unsophisticated foods. The mortality of the poorly fed animals was 34.1 per cent whilst that of the controls was below 1 per cent. The experimental animals were far more susceptible to infections. In general, diseases of the lungs and the gastro-intestinal systems were the most common. Other illnesses included night blindness and xerophthalmia, dental caries, polyneuritis, beriberi, pellagra, inflammation of the mouth, diarrhoea, scurvy, rickets, osteoporosis, slow healing of fractures, sterility, stone in the bladder, anaemias and some types of goitre: in fact, common symptoms of the diseases of humans eating similarly faulty diets. He also recorded irritability, listlessness and violence towards other animals. People tended to disbelieve these behavioural effects, but many experimenters report that one of the signs of poor nutrition in a colony of rats is cannibalism of the newborn by the mother. This kind of data seldom reaches the published literature because of fear of embarrassing the experimenter. It is, however, interesting.

Always McCarrison's controls fared far better and, as he pointed out: 'All these illnesses were brought about directly or indirectly by faulty diets – admittedly some very faulty – in use by humans.' It was therefore reasonable to suppose that humans would suffer in the same way. McCarrison concluded from his 20 years' work in India:

> Malnutrition is the chief cause of lowered resistance to infection . . . the reason why [the Indians] succumb by the hundreds of thousands to the ravages of such scourges as malaria, kalazar, cholera, dysentry, leprosy and tuberculosis . . . Perfectly constituted food is not a panacea for all diseases, but it is an agent as potent in preventing a host of them as is the mosquito net in preventing one, or inoculation in preventing another (McCarrison, 1953).

McCarrison founded the first link between medicine and nutrition research by creating a Nutrition Research Centre in Hyderabad. He described, at the beginning of the century, some of the effects of departing from the pattern of nutrition that is natural for the species *Homo sapiens*. In 1966 a society, the McCarrison Society, has been established in Britain by doctors, dentists, nutritionists and scientists to honour his name, work and message.

Another pioneer of about the same time was an Amercian dentist named Weston A. Price who in his middle age closed the door of his surgery and went sailing round the world for a number of years, comparing the teeth of primitive people living in isolated communities with those of people who had previously been isolated but had recently been reached by the 'developed' world and its influence. What drove him on this long voyage was his research on dental degeneracy. He needed to find communities free from dental decay to use as controls and nowhere in the modern world, neither in the USA nor in the rest of Europe, could he find them.

Everywhere he went he made a most thorough study of teeth, bone-structure and diet. He looked at every tooth in the head of every member of the communities he visited and also examined each one's jaws and dental arch (the upper jaw and bony structure forming the roof of the mouth). He took samples of the food they ate, and the soil it grew in, and sent it back for analysis, and he wrote up the results in a book called *Nutrition and Physcial Degeneration*, first published in 1938. His conclusion was that people living in the old cultures had excellent teeth, a fact known to historians, and he expressed his belief that 'there had been more dental decay in the last hundred years than there had been in any previous millennium' (Price, 1938).

Price's travels took him across the world from Switzerland to the Amazon. He studied primitive and modernised communities in Alaska, American Indians in Canada and various parts of the USA, inhabitants of the coastal and mountainous regions of Peru, Melanesians and Polynesians in the archipelagos of the South Pacific tribes in eastern and central Africa, aboriginals in Australia, Malaya, and on the islands north of Australia, and Maoris in New Zealand. Where possible he also studied white people living in or near these communities. His most striking and saddening findings concerned the primitive peoples who had been evicted from their traditional homes by the white man and settled on inferior lands, where they ceased to grow or find their old foods and lived instead on shop-bought European produce. No doubt white bread, refined cereals, sugar, jam and tinned fruit seemed tasty and colourful especially if plied with alcohol to distort your vision of the world: nutritionally the change was disastrous. Often Price came across Europeans in charge of such populations who were extremely distressed at the worsening health of the people under their care. The agony of the natives themselves as they saw their children visibly deteriorate makes very sad reading.

To Price, the most directly obvious effect was bad teeth. Less obviously there was a common shrinking of the jaw and dental arch which meant that the same number of teeth had to come out of a shorter space so that some of them protruded from the sides of the jaw, as can be seen in Westerners. At the time it was common wisdom that such deformities arose from the crossing of pure races, but Price found that in reality pure races suffered from the same troubles. He set out to discover the true causes of tooth decay and established that it was directly controlled by nutrition. The shrinking of the jaw arch was caused by lack of minerals, not enough to produce the dramatic effects of acute vitamin deficiency such as scurvy or beriberi but enough to cause physical change. His results are somewhat parallel to those of Professor Donald Gebbie's studies on the pelvis where he sees malformation and disproportion equated with poor diets.

The photographs in Price's book of the faces of people from the same race but now living on different foods (some good, some bad for teeth and jaws) in different places, are an extraordinary witness to the effect of nutrition not only on dental health but its actual effect on the shape of the face itself. Seeing his pages of photographs would persuade the casual observer to believe that not only had food affected shape, but that colonies of people now looked different from one another whereas before, within a colony, they had all had a familiar likeness.

Such geographical differences can today be seen in action in Japan. Twenty years ago the total fat intake was about 10 per cent of dietary energy. Today it has, under Western influence, risen to 27 per cent. This with other changes (namely the introduction of milk, meat, butter and margarine) has led to an increase in height and changes in body and facial shape, so that many who have been wooed by Western diets are beginning to look like Westerners – a contrast seen by looking at the faces of 30-year-old compared with 60-year-old men. With these changes in shape and face is an increase in incidence of typically Western disease.

The results of Price's elaborate research on the teeth of many hundreds of people are vividly displayed by the following table, which shows the percentages of teeth attacked by caries in primitive and modernised groups of the same peoples:

Table 4 Percentages of teeth attacked by decay in different peoples

	primitive	*modernised*
Swiss	4.6	29.8
Gaelics	1.20	30.0
Eskimoes	0.09	13.0
Northern Indians	0.16	21.5
Seminole Indians	4.0	40.0
Melanesians	0.38	29.0
Polynesians	0.32	21.9
Africans	0.20	6.8
Australian Aborigines	0.00	70.9
New Zealand Maori	0.01	55.3
Malays	0.09	20.6
Coastal Peruvians	0.04	40.0+
High Andes Indians	0.00	40.0+
Amazon Jungle Indians	0.00	40.0+

The appearance of the older people often conformed to a clear racial type and Price has several photographs in his book showing how strong that type could be, to the extent that members of same tribe or group often looked like brothers or sisters, even though they were not related. He considered that these differences could be seen in people from the same stock but living in different islands, with different foods. These differences were especially marked in comparing fish eating with agricultural people.

Price's own conviction was that the changes were associated with a change in genetic structure. He wrote: 'The forces involved in heredity have in general been deemed so powerful as to be able to resist all impacts and changes in environment. These data will indicate that much that we have interpreted as being due to heredity, is really the result of intercepted heredity.'

Serious dental degeneration occurred in the first generation after the introduction of modern food to the *parents*, as did the change in facial structure. Price believed that the teeth and jaws were not the only things affected by poor nutrition. He thought that it could cause deterioration of various kinds, including hare-lip, cleft palate, narrow hips, narrow face, constricted nostrils and defects of the skull showing in the face and the floor of the brain. All of these factors were in his view associated with deficiencies of vitamins and minerals both before and during pregnancy.

He also believed that the same was true of brain defects and mental retardation and considered that mental states could be strongly affected by sub-optimal nutrition, particularly a shortgage of vitamin A and the B vitamins, which are needed to maintain the structure of the nervous system. Price himself commented:

> The frequency among juvenile delinquents of bodily weakness and ill health has been remarked on by almost every recent writer. In my own series of cases nearly 70 per cent were suffering from such defects . . . If, as now seems indicated, mal-development with its production of physical, mental and moral cripples is the result of forces that could have been reduced or prevented, by what program shall we proceed to accomplish this reduction or prevention? (Weston Price, 1938)

This insight fits closely with modern scientific thought.

His own practical answer had been to run a food mission for deprived children during the depression of the early 1930s. Price's work was ahead of its time and should be reassessed. It gives us another example of the way in which a change in diet from baseline to modern can influence anatomy and health. His tribute to the peoples he studied is well worth remembering. Commenting on the health of those whose diets had escaped the modern influence he said: 'Our modern civilisations are doubly indebted to the primitive races for they have both demonstrated what we might be like in physical form and health, and have indicated the nutritional requirements for doing so (Price, 1938).'

McCarrison had pointed out the danger of moving from unsophisticated foods to refined rice in India. Price had shown how the same movement to refined carbohydrates had an adverse effect on teeth and jaw structure in many parts of the world. Donald Gebbie has described how similar developments from our common nutritional baseline influenced the pelvis and birth in many developing countries.

One researcher who recognised the effect of this change from the natural balance to 'purified' food in Western society and made this the centre of his work in the 1950s and 1960s was a ship's surgeon, Captain T. L. Cleave.

Cleave's ideas were also 'based on human evolution, and stemming from it the perfect adaptation, universal in extent and profound in degree, that excited Darwin from an early date [on which] his theory of evolution was primarily based.'

Cleave corresponded with doctors and hospitals all round the world to get information on the geographical distribution of diseases and found that Western diseases were far less common in the Third World. This, he believed, could not be a matter of genetic differences because the degenerative diseases affected far more people than would be expected if they were genetically determined. Defects that are known to be inherited, such as cleft palate, harelip, congenital malformations of the heart and club foot, affect at most five people per thousand. Varicose veins, on the other hand, affect 100 people per thousand and peptic ulcer, 200.

Cleave thought that many of the troubles of Western man should be seen as symptoms of a single disease – the 'saccharin disease' as he called it, meaning that it was related to sugar. The 'symptoms', or individual diseases, included diabetes, diverticular disease (in which little pockets form in the lining of the colon and fill with stagnant, putrefying material), obesity, peptic ulcer, constipation and varicose veins. An important point was that each of the individual diseases had its own incubation period. Decay in children's milk teeth could start to show in a few months but diabetes might not become a problem until 20 years after a community began to eat large amounts of sugar and refined carbohydrates. Diverticular disease could take 40 years to show in a population.

The chief culprits, Cleave believed, were refined carbohydrates, refined cereals, refined sugar and alcohol (which is produced from sugars). The damage was caused by three factors: the reduction in fibre, the overconcentration of high-energy food, and the removal of significant amounts of essential nutrients. It was not the presence of sugar or carbohydrates that caused the trouble: it was their isolation from the other things that accompanied them in their natural forms: again a move away from unsophisticated foods.

As an example, Cleave cited the Indians of Natal, who were eating 60 pounds of sugar a year compared to 6 pounds for Indians in India. (The figure for Britain in 1960 was 120 pounds.) In Natal there was a high incidence of diabetes, whereas in India the incidence was very low, and of a different form. Even more significantly, the Zulu and Pondo cane-cutters in Natal, who were allowed to chew as much as they liked, were 'singularly clear of the disease'.

Cleave's mistake was in being too inclusive and giving the impression that all degenerative diseases were a result of eating sugar or refined carbohydrates. For example he considered fat had nothing to do with heart disease. Whilst researchers in heart disease were willing to give Cleave a point or two, he was intransigent. This was forgivable: the knowledge we now have on the nutritional value of essential fats came after his time. His views caused a lot of scepticism but there is a growing realisation that the importance of his work was underestimated.

One important researcher who was impressed by 'the undeniable truth and logic' of Cleave's idea was Dennis Burkitt, a surgeon who worked for many years in Uganda, had a thorough knowledge of Third World patterns of disease, and was aware that Western disease rarely occurred in tribal Africans, though they did start to appear when the Africans moved into towns and cities.

Burkitt's great contribution was explaining Cleave's understanding of the importance of fibre. He pointed out that food goes through the digestive system very much faster when the diet contains plenty of fibre: dense, fibreless foods cause 'sluggish gut', constipation and a host of other diseases which are all related to the stagnant food passing through the intestine at a rate far slower than nature intended. Burkitt supported Cleave's hypothesis that tumours of the bowel, 'affecting at least a quarter of the population over fifty', appendicitis, gallstones, diverticular disease, hiatus hernia, varicose veins, haemorrhoids and obesity, as well as simple constipation, are all connected with lack of fibre in the diet. The sluggishness of movement through the bowel can be cured by simply adding fibre. The fibre in fruits, vegetables and brans is very effective (although brans can easily be overused: excess bran can interfere with the absorption of iron and zinc, leading to deficiencies). Thanks to the work of Cleave and Burkitt the role of fibre is generally accepted and a high-fibre diet is used in many hospitals as part of the treatment of the illnesses concerned.

In fact, Cleave and Burkitt, McCarrison and Price were spelling out the dangers of drifting from our baseline foods. Professor John

Yudkin, in his *Pure White and Deadly* (Yudkin, 1972), and other publications has been warning us of similar dangers from overconsumption of sugar.

It was tumours of the bowel, the most frightening of the diseases listed by Burkitt, that attracted the most attention. To many people the idea that cancer was connected with food came as a disturbing surprise, and the fact that one kind of cancer could be largely prevented by simply adding fibre to one's diet was revolutionary.

When Sir Richard Doll made a study of a wide range of cancers and their incidence in different countries and published his results in 1968, he concluded that the discrepancy of different types of cancers occurring in different communities was so wide that something between 30 and 70 per cent of all cancers could be related to food (Doll, 1968). At the time this proposition was submerged in the wave of excitement and controversy over his other great discovery, the connection of lung cancer with smoking. There is, however, substantial evidence that food (chemicals) and cancer are intimately connected in a number of ways.

Once again, Africa provides some interesting lessons. East Africa, and particularly Uganda, is a fertile area for research into nutrition. Uganda is a country of great contrasts: here are the snow-capped Mountains of the Moon, the crater lakes of the Queen Elizabeth National Park, the Semiliki Plains, moving with the gold of the Uganda kob, and the waters of Lake Albert with its giant Nile perch. Then there are the impenetrable forests of Kabale with dense undergrowth and high canopies out of which the steam clouds rise to moisten the morning air. These contrast with the semi-arid plains of Karamoja where giraffe and eland thrive on the deep-rooted acacia and balanites trees even in the burnt landscape of the dry season. Around Lake Victoria the vivid greenness, the equatorial rain, sunshine and soil combine to create what is perhaps the greatest biological productivity in the world.

The food produced today in these contrasting places naturally varies a great deal. On the shores of the lakes fish is eaten fresh and surpluses dried and sold. In the semi-arid north-east the people live on a mixture of hunting, nomadic cattle husbandry and seasonal seed crops like millet and sim-sim. In the areas around the lakes they raise crops of ground-nuts, sweet potatoes, cassava, yams, maize and some 32 varieties of bananas for cooking and as fruit. These plants are *not* indigenous to Africa. They were brought over from South America, when the trade routes, and then the slave trade were opened by shipping. Maize and ground-nuts were needed to feed up the slaves before they were transported.

These plants proved so productive in their new home that they soon ousted the traditional foods won by hunting, gathering and fishing. It is said that if you plant a banana tree on the day a child is born, it, with its many offshoots, will bear enough fruit to feed him for the whole of his life. Prior to these changes, cattle and goats had come in from the more temperate zones of the north of Africa. Unsuited to the hot, dry savannah and bush they led their human supervisors a merry dance in the nomadic search for water. Long-thorned bushes and trees that were once food for eland and giraffe were felled to barricade the cattle at night and protect them from lions, turned into charcoal for cooking and burnt out to clear the way for the fresh green grass so loved by cows. When the green flush was over and the grass died in the burning heat of the sun, they moved on, leaving dust behind.

As the new foods and new ways of life took over, the human population increased. The wild animals and many of the wild plants were forced out by the new crops and today, over large areas, the plantain, banana, sweet potato and cassava plantations stretch almost as far as the eye can see. The range of food the people now eat is consequently much narrower. In Uganda, for example, the Baganda people may live very largely on anything up to 2kg of cooked plantains a day.

If the change from wild and unsophisticated foods on the Indian continent had led to vitamin A blindness, pellagra, beriberi, and the related disorders described by McCarrison, what happened in Africa?

Two major differences can be seen in different African peoples: form (shape and size) and disease. The peoples like the Baganda and the Bahutu who live on plantains, sweet potatoes and other such crops are short and round, while the cattle people, such as the Karamajong, the Masai, the Acholi and the Samburu are tall and slender. Indeed, the Watutsi and the Bahutu lived together in Burundi. The Watutsi were the masters and the Bahutu the servants. The Bahutu managed the cattle and tended the gardens but the milk and meat was reserved for the Watutsi and the Bahutu had the sweet potatoes. The Watutsi, men and women alike, were tall, slender and graceful whereas the Bahutu were squat, dumpy fellows.

Of course the difference in stature of the Watutsi and the Bahutu and the Baganda may be genetic. However, the Kabaka, the King of Baganda, although not tall, was none the less more like a smaller version of the Watutsi. One wonders if these differences are Africa's version of our socio-economic differences. In the UK in general, the lower socio-economic groups are smaller, fatter, have a higher incidence of obesity, a greater risk of modern diseases and shorter life spans. These differences of form are most likely to be due to different gene expression responses to the contrasting nature of the foods eaten rather than any fundamental genetic difference.

This data cross references strongly with the evidence of Weston Price and it is difficult to escape the view that food is relevant to the shape and physique of the different tribes. The El Molo provided clearer evidence. Here, we have a people with curved legs living exclusively on the produce of Lake Rudolph and the Jade Sea, on a low calcium/high phosphorus diet. Their legs were not curved for reasons of differences in genetics, but as a direct result of their diet.

Although dependence on lake produce had distorted the anatomy of the El Molo, they were otherwise happy and healthy. Others were not so healthy. While bananas and plantains contain a number of nutrients, they do not contain them all. In particular they are short of protein and fat-soluble nutrients. Kwashiorkor, a serious form of infant malnutrition, is endemic amongst the people who depend on them. It affects the firstborn not long after the second child is born. Europeans thought it to be caused by congenital syphilis but Dr Cecily Williams, working in Ghana, recognised it as nutritional. Infants were being weaned on to these high carbohydrate, low protein, low fat foods: protein/calorie malnutrition followed. Although food shortages affected the cattle people and their children suffered from marasmus, kwashiorkor, with its oedema, red thin hair and pot belly, was relatively uncommon. It was particularly prevalent amongst the plantain–sweet potato peoples.

In East Africa there is a further disease found in an unusually high frequency in specific geographical or tribal locations. It is known as volvulus of the sigmoid colon, a sort of kink in the intestine which can block it, cause it to rupture and turn gangrenous, so leading to an agonising death. Volvulus and even double volvulus was the commonest surgical emergency in plantain communities. It is seldom seen in Europe, nor in the Karamajong, the Masai or the hunters and gatherers. It appears that this disease is closely related to a diet of plantains and sweet potatoes, for where European diets produce sluggish movement through the gut, plantains do just the opposite. For a start, two kilograms of steamed plantain a day means a lot of bulk to be passed through and plantains contain plenty of material that is not digested. They also contain unusual amounts of serotonin, a hormone-like substance which stimulates the movement of the intestine. The bulk and the speed of transit could be too much for the gut, which follows a tortuous course at the best of times. The likely explanation of the volvulus is a poorly nourished intestinal muscle, combined with the very fast passage of food.

Mount Elgon rises 10,000 ft above the plains of Karamoja. Its slopes are covered in trees and exotic plants which enjoy the mountain mists and rivers created by its own microclimate. The people who inhabit these slopes grow plantains and attend the hospital at Mbale at the foot of the mountains. The people of the plains, who enjoy the river course of Loitokitok, also attend the same hospital; they have no volvulus, the people of Elgon do.

Whilst the Baganda did not experience our form of heart disease, they suffered from a different form whereby, instead of having furred up arteries, the muscle and the inside lining of the heart breaks down: muscle is replaced by scar tissue and people die from heart failure. It is called endomyocardial fibrosis (EMF) and was described by Professor Jack Davies in Uganda. Professor A. G. Shaper of the Makerere Medical School was interested in EMF as he wanted to know why it was that Americans had coronary disease and Africans did not but had EMF instead. In the early 1960s Professor Shaper reported that the African adults had low blood cholesterol levels whilst the Americans had high levels. However, he also pointed out that EMF was most common in the low socio-economic groups and was particularly prevalent amongst the plantain eaters, especially in women after the birth of a child. This latter piece of evidence suggested a nutritional background related to undernutrition. Experimental studies on guinea-pigs and dogs fed low protein diets reproduced a similar damage.

It may be an oversimplification to suggest that in Uganda it is the heart muscle which is at risk because of its dependence on protein, whereas in Europe and the USA it is the arteries that are at risk because of their requirement for lipids. It is very likely that the different nutritional background is the reason for the different types of heart disease.

The ground-nut, another new food foreign to Africa, is also associated with an unusually high frequency of disease: this time primary liver cancer. In Europe primary liver cancer is rare. Liver cancer may occur but it is usually secondary to alcoholic cirrhosis.

In the UK during 1960 there was an outbreak of the disease which killed thousands of turkeys. The Ministry of Agriculture's laboratory quickly isolated a toxin and pin-pointed its source to mouldy ground-nuts. The compound was called aflatoxin and turned out to be one of the most powerful carcinogens known. It was especially carcinogenic in the liver. A paper from Uganda (Lopez & Crawford, 1967) showed that the toxin was present in the ground-nuts sold for human consumption in the local markets at levels which would offer an explanation for the high incidence of liver cancer. Much more is now known about aflatoxin and other fungal toxins which result from poor food storage. Some still believe that the high incidence of primary liver cancer is due to hepatitis and it is always difficult to prove these matters. What about a combination of the most powerful liver carcinogen known, hepatitis and a poor state of nutrition?

There is another interesting discrepancy in disease that appears to be food related: a higher incidence of bladder cancer occurs among the banana eating peoples compared with others eating different foods. The reason may be that the banana is particularly rich in indoles which the body converts to 3-hydroxyanthranilic acid en route to other more useful components. Strangely enough, although this substance is made in the body, it is a carcinogen and belongs to the group of compounds identified as causing the high incidence of bladder cancer in workers from the aniline dye and rubber industries. Dr Boyland of the Chester Beatty Cancer Research Institute in London proved in experimental animals that these compounds were carcinogenic. Steps were then taken to stop them from contaminating the skin and so getting into food, and the bladder cancer vanished. In some countries, the industry did not take protective action and the incidence of bladder cancer remained high in their workers until they did.

There are two interesting points about this story. The first is the fact that the body makes carcinogens from other non-carcinogenic compounds. Indeed, the indole nucleus is part of the essential amino-acid tryptophan which is an essential component of body proteins and food proteins. Hence it is even worse because this carcinogen is being synthesised from a nutrient that is essential for animal life. Normally, it is present in the urine in only trace amounts and is of little concern but the studies in East Africa showed that its concentration in the urine of the plantain eaters was many times that of those who lived on sorghum, maize or millet.

The second point is the site at which the cancer occurs. This carcinogen can produce a tumour at any point in the body at which it concentrates. An implant under the skin will produce a tumour at that location. There is nothing specific about 3-hydroxyanthranilic acid which causes cancer in the bladder rather than elsewhere. What happens is that the kidneys concentrate it in the urine, keeping the rest of the body clear and putting the bladder at risk.

These same African communities did not have baths and showers. Unless water was readily available because they lived on a lakeshore or beside a river, they had to carry the water from wells, rivers or other supplies. Hence water was not the sort of thing to shower yourself with daily. However, when a circumcised man did wash his body, the head of the penis would be cleansed especially if he immersed himself in water. By contrast, the head of the penis of the uncircumcised man would not be washed free of its carcinogens, which would become even more concentrated under the foreskin, possibly for long enough to induce cancer.

It may just be an extraordinary coincidence, but Dennis Burkitt quite independently reported a remarkably high incidence of cancer of the penis in the uncircumcised people, particularly the plantain eaters. It was uncommon in the circumcised and non-plantain eaters. Bladder cancer in East Africa is a relatively minor concern. However, in the uncircumcised plantain eaters, cancer of the penis is virtually top of the cancer incidence league: somewhat equivalent to breast or colon cancer in Europe and North America. In the high spot areas, cancer of the penis accounts for about 25 per cent of all cancers compared to only 4 or 5 per cent in neighbouring Kenya.

If bladder cancer can be explained by a carcinogen being concentrated in the urine, why is aflatoxin specific to the liver? Again, there is a simple chemical answer. In the blood, aflatoxin is bound tightly to plasma protein and is therefore not free to be excreted by the kidney so other tissues are protected. On the other hand, the liver is responsible for metabolising blood proteins, it strips aflatoxin off the protein and excretes this carcinogen in the bile. Hence aflatoxin appears in the liver in its naked state, able to attack the cell's DNA and cause cancer. Tumours do, of course, occur elsewhere, but the liver is the principle site. By contrast, the 3-hydroxyanthranilic acid is actively excreted by the kidney which keeps levels low in the blood, but concentrates it in the bladder. The conclusion one draws is that the target organ is decided by the physiological handling of carcinogens.

So here we have evidence of food and lifestyle deciding disease patterns. What is most interesting is that the diseases we have talked about are related to foods which are *new* to Africa: primary carcinoma of the liver, volvulus and double volvulus, kwashiorkor and endomyocardial fibrosis. If the suggestion on cancer of the bladder and penis is correct these can also be added. These diseases have no place in Europe or the USA and are of little concern to the Hadza, the hunters and gatherers of East Africa or the cattle people. They should have no place in the rest of Africa either.

A similar history could be written by others about the Far East, the Maoris, the Australian Aborigines or the Polynesians on the Island of Nauru. However, this story of Africa is written with first-hand experience and can be simplified as follows: it was the trade routes that brought kwashiorkor, volvulus and liver cancer to Africa. The introduction of new foods and new agricultures created new disease in Africa. It is not so easy to admit that we in Europe have made the same mistake. The foods we eat today in Northern Europe and North America are yet again different. They have been changed in different ways: so have our diseases.

12 European departures

Of all the changes which have taken place in moving from the baseline, unsophisticated foods of nature to what we eat today, the European manipulation of our domestic animals has been one of the most illustrative. It is a striking example because the change has been so large and because its new end product has been simulated in so many different ways which today we accept without question.

There is no doubt that today's domestic animals are different from their wild counterparts. Once man began to selectively breed animals and determine what they ate in order to favour this or that useful characteristic, he also began to introduce unintended changes. Juliet Clutton-Brock has some interesting comments to make on this in her book *Domesticated Animals from Early Times*. She points out that domestication 'causes an imbalance in the rate of growth of different parts of the organism' so that the adult animal is shaped differently from its wild counterparts:

> Within a very few generations of breeding in captivity, the facial region of the skull and the jaws becomes shortened, this being common in many species but is most apparent in early domestic dogs. At first there is no corresponding reduction in size of the cheek teeth which are genetically much more stable than the bones of the skull. This causes a crowding or compaction of the pre-molars and molars, a character that is used to distinguish the remains of the earliest domestic dogs from those of wild wolves (Clutton-Brock, 1981).

Her description of these changes reminds one of Weston Price, writing on the changes in form of primitive people introduced to modern Western diets.

The sound-case of the ear drum (known as the timpanic bolla) is much smaller in a dog's skull than in a wolf's; the same thing can be seen by comparing the skull of any domestic species with the corresponding wild one. The timpanic bolla of a sheep's skull seems like a vestigial relic when compared with that of a wild species of similar size. This sense organ seems to become smaller in domestic animals relative to their overall size, and one of the most significant changes is that their relative brain size also shrinks. The reduction is substantial, around 30 per cent.

This loss in brain and sensory capacity is additional testimony for the idea presented in the chapter on *Homo aquaticus* that body size can increase while at the same time relative brain size can diminish. Furthermore, it is tempting to make a connection between this loss of brain capacity and another change which is perhaps the most significant of all in its effect on the humans who eat domesticated animals. It is so unexpected that when first discovered, it took years of argument and a series of scientific papers to convince the sceptics. The story starts with a buffalo shot in Africa. The animal looked well fleshed and fat and was shot for that reason to provide meat.

When the buffalo was skinned, the most striking feature was the lean nature of the meat: there was virtually no fat to be seen. There was no question of the animal being short of food: it provided mountains of red meat. The picture many people have of wild animals living precariously near the edge of starvation is wide of the mark; this particular buffalo had spent its life in the most lush part of Uganda where tropical heat and ample water produce grass in abundance. These riverine areas of East Africa have an enormous primary productivity, up to 60 tons of plant life per acre in a year. Certainly the buffalo was not lean because it was short of food.

A second unexpected difference became apparent when the buffalo meat was analysed to see if its fatty acids were any different from English beef. Animal fats were assumed to be the rigid saturated type, such as is found in domesticated 'fatstock' and used to make candles. This 'fact' has led several Government committees to recommend eating less meat to reduce heart disease on the assumption that meat fat was the same saturated fat.

However, this assumption turned out to be incorrect. Instead of finding the buffalo meat full of the saturated fatty acids as expected, it was rich in the essential polyunsaturated fatty acids. Meat samples collected from other buffaloes and wild herbivores were all, without exception, rich in essential polyunsaturated fatty acids: the very opposite of the accepted view of meat as a saturated food to be avoided.

Up to this time it had been generally believed that the essential structural lipids were destroyed in the stomachs of ruminant animals. Buffaloes are ruminants so the idea that buffalo fat contained a high proportion of essential, polyunsaturated lipids was automatically greeted with disbelief. A piece of the buffalo meat was then given to the laboaratory of the Government Chemist, where it was subjected to their analytical techniques. They confirmed and published the fact that it was indeed rich in linoleic and alpha-linolenic acids, the two parent essential fatty acids used in structural lipids.

When the article dealing with this appeared in the *Lancet* in 1968 it attracted much criticism. The analytical methods had to be wrong. Some ten scientific papers later, at least some acknowledged that the data was correct. The reason for the confusion was the use of the term 'animal fat'. As described previously, there are two types of fat in the body: (i) storage; and (ii) structural. The storage fat is in a sense the rubbish dump reserved for burning as fuel. The structural fat is tailored to the requirement of cellular function and it is therefore built with essential polyunsaturated fatty acids – and quite different in nature to the depot fats.

If an animal is overfed and denied exercise it simply gets fat. If the process is continued long enough, it loses muscle, and depot fats infiltrate the retreating muscle fibres, giving the 'marbled' appearance characteristic of modern intensively fed animals. This infiltrating fat is, like the rest of the storage fat of ruminants, saturated fat. Analysis of meat from this type of animal shows its fat to be largely saturated.

By contrast, under natural or free-living circumstances this cannot happen. The animal eats the right kind of food in amounts appropriate to its growth and exercise and the requirements for the different seasons. Hence analysis of its meat shows it to be characterised by the type of fatty acids involved in cell function and structure: that is, it is rich in the essential fatty acids.

Here in the buffalo meat was plain evidence that throughout man's evolution, the animals he ate would have had a low fat content: and what fat they had, contained a high proportion of essential structural lipid – quite different to 'marbled' beef. The fat in the domestic animal was the opposite: importantly, it is the kind of fat that experimenters had linked with heart disease. Man had unwittingly changed his animals in a direction now known to increase the risk of heart disease.

What went wrong to produce such an unhealthy result? The

answer emerges when the history of the development of modern meat animals is examined: in summary, they have been changed by alterations in the conditions under which the animals themselves were made to live.

> The natural foodstuff of the ancestor to our domestic cattle was soft bushy, leafy material, the lower branches of trees, sedges, herbs and grasses. This made it ideally suitable for operating in the forest openings and encouragement of the numbers of these species would undoubtedly have led to a wider and more extensive use of materials. Paintings of cattle done in the seventeenth, eighteenth and nineteenth centuries show them in very much this kind of environment. The vegetation is clearly mixed with trees, bushes and grasses and contrasts starkly with the present use of open fields and electric fences. Darwin provides incidental confirmation of this situation when he discusses the evolution of the long neck. He asks the reader to look at a field of cattle where he will see the lower branches of the trees have been 'planed to an exact height' where the heads of the cattle can reach (Crawford & Crawford, 1972).

Before the enclosures of the 17th century, cattle had mostly been herded in open grass, bushland and forest. When people began to contain them in fields, these were at first fairly simple affairs. They were not cleared and they did not have special grasses growing in them. Even so, the variety of food that the animals could select for themselves was drastically reduced. They ate the hedgerows, but the human response to that was to replace the succulent bushes with thorns. Enclosing an animal in a field meant simply that man now decided what the animal should eat.

The next move was intensification. A paradox in animal management arose fairly soon after the enclosures were created. When the number of animals in a field was increased, instead of doing worse as the food supply was shared out they did better. The explanation of this paradox is simple. If grass is allowed to grow up unchecked it produces tall stems with the seed heads at the top. The stems are rich in fibre but not in energy. On the other hand, if the grass is kept cropped – either mown or grazed – it pushes out young, fresh tips which are very rich in energy. Intensifying the number of animals on a field therefore increased the energy of the animal's food throughout the growing season. The simultaneous reduction in the amount of exercise reduced muscle development and led to more fat deposition. This tactic reached its zenith in the more modern use of stalls in which to keep and feed the animals. Added to all this was the clever idea of castration, again to make the males quiet and to gain weight faster: but the weight was largely fat – saturated, storage fat. When people later became technically competent enough to manufacture feeds for animals, they modelled the energy balance not on the autumn but on the high-energy spring grass. Hence the development of 'high-energy' feeds which were heralded with the advertising claim that they put on weight much faster than before.

Many know from experience that when anyone puts on weight rapidly what he or she puts on is storage fat, and the same is true of animals. The high-energy grasses made the production of fat carcasses easy and profitable. The process was pushed a long step further by the introduction of high-energy winter feeds. In *As You Like It* Shakespeare's shepherd remarks how his sheep grow fat in the spring and lean in the winter, and these seasonal variations are a necessary part of biology. Modern farming methods force animals to grow steadily fatter all the year long. They live on a perpetual spring diet.

This high-energy diet combined well with another development. When the animals were sold in the market, the heavier they were the better the price. So they then chose those that put on weight fastest for breeding with the aim of producing big, round well-covered animals. They fell into the trap of selecting genetically for the fat animal. The combination of all these tactics culminated in the prize-winning beast at the Smithfield Show a few years ago. The butcher who bought it, hoping to win credit for his company by selling this highly acclaimed carcass, found instead that the meat was unsaleable: it was so fat that even in England no one wanted to eat it, or so the story goes.

Mr Gordon Williams farmed animals in the hills above the Wye valley in Wales for most of his life until he retired in 1971. His farmhouse has a bread oven in the kitchen wall and he described how they cooked large amounts of meat for the villagers who helped cut the hay in the autumn. They also used the bread oven to make candles out of the white storage fat from their sheep and cattle to light the house during the winter. Electricity did not reach his farm until the late 1940s.

There are two interesting points which emerge from Mr Williams's story. First, the amount of exercise involved in getting to and from work, in work itself or even cutting enough wood to keep the family warm in the winter would have burnt up many calories which today are conserved because we drive by car or bus and have central heating, cooking fuel, electric light, and hot water on tap. The second point is that without electricity the house had to be lit by candles. Without Kuwaiti, Texan or North Sea oil, those candles were made from the white storage fat of beef and sheep. Beeswax provided an additional material for making candles but there were obviously only a few of these and they were most used in churches. Today candles are made from paraffin wax obtained from the oil companies.

Initially storage fat would have been an important form of energy, for people had to work hard and the winter months would have added an extra demand in energy expenditure. However, the fat was not only used for food, but also made these candles and boot and saddle polish. Today, with our sedentary lifestyle, the high fat carcass is an anachronism. These commodities are made from fossil oils and the fat now goes into the manufacture of pies, sausages, pastries, cooking fats, margarines, ice-cream, biscuits, cakes, convenience foods, TV dinners, quick-chill meals, snack foods, crisps and even bread.

What is the size of the fat production problem? In the UK the present meat consumption is about 3.9 million tons per year. The energy contained in its storage fat represents about 1.097 times 10 to the power of 14 Joules a day, or enough to keep an oil-fired 1200 Megawatt station in operation for a year. Translated into candle power, the present animal production in the UK provides enough white storage fat for all families to throw away their electric light bulbs. We now eat the candles.

Let us look at the figures. The Meat and Livestock Commission in Britain is trying to persuade farmers to aim at 25 per cent carcass fat, but even if that aim is achieved, the comparisons with wild animals are heavily to the disadvantage of the domestic breeds. A carcass with 25 per cent fat would carry about 50 per cent lean meat. When Dr Ledger dissected over 220 animals from 16 different wild species he found that the average amount of fat was around 3 or 4 per cent and that of lean over 75 per cent: a large difference.

The difference is even more significant if we allow for the fact that most of muscle is water: 80 per cent of it! Take that away and look at the amount of protein and nutrients compared to the amount of storage fat (see Table 5).

Table 5 Comparison of proportion of lean meat and nutrients in the carcasses of wild and domestic beef animals (%)

	wild	*domestic*
lean meat	75%	50%
protein & nutrients (water removed)	15%	10%
storage fat	4%	25%

If we compare the *calories*, or energy, we get from the two sources – wild and domestic – the figures become still more interesting. Just reading the actual amounts of protein, carbohydrate or fat on a packet is misleading because the contributions to dietary energy by protein and fat or carbohydrate are quite different. Protein produces only 4 calories per gram whilst a gram of fat provides 9 calories. Converting the lean (20 per cent of which is protein) and fat (which is virtually all fat) from wild and domestic beef using the data in the table above gives the following:

Table 6 Comparison of calories obtained per 100g fat/lean

	fat	*lean*	*ratio fat:lean*
wild animal	36	70	0.5:1
domestic animal	225	40	5.6:1

The above table illustrates just how far the development of the fat animal has gone. It has happened little by little, but the end result is that we are eating obese animals. Although people say that they rear animals to get protein, they are actually producing far more fat than protein. In domestic breeds the fat provides more than five times as many calories as the lean. The differences between the wild and domestic animals are staggering. One thing is certain: we cannot blame the animal for it.

Even that is not the end of the story. What is termed lean meat in butchers' carcasses is not lean meat at all. The tissue is infiltrated with veins of fat which can account for anything up to 20 per cent of its weight. In wild animals there is virtually no visible fat between the muscle fibres. The only facts present are the structural (essential polyunsaturated) lipids used for building the cells. Any surplus fat the animals have is stored around the interior organs such as the kidneys and heart. People have selectively bred overweight animals, unhealthy in themselves and unhealthy to eat.

What is true of cattle is also true of other domestic species. In medieval times pigs roamed the forests and during the winter were fed largely on acorns gathered from the oaks. It happens that acorns are particularly rich in polyunsaturated essential fatty acids. About 50 per cent of their energy is contained in an oil, about half of which is linoleic acid with 9 to 12 per cent alpha-linolenic acid. Both parent essential fatty acids are there in plenty, together with protein, vitamins B and E and other essential nutrients. The effect of this diet can be seen from the comparison of wild and modern domestic pigs. The fat in and around their muscles amounts to no more than about 2 per cent of the muscle weight, and in their bodies as a whole, the ratio of essential polyunsaturated to saturated fats is about 2 to 1. A modern pig fed on a high-energy diet produces a pork chop with 40 to 80 per cent of its energy as fat and a ratio of polyunsaturated to saturated fat of only 0.2 to 1. In that ratio the wild pig does ten times as well. Its meat is also superb.

Even chickens, which traditionally were very low in fat, were brought into line. In its 1976 report, *Diet and the Prevention of Heart Disease*, the Royal College of Physicians urged people to eat chicken in place of red meat because it was low in fat. The Royal College reflected a popular misconception which is now out of date. At the end of the last century the carcass fat on chickens was indeed a mere 2.4 per cent* but by the end of the early 1970s the data reported by the Agricultural Research Council Centres showed it

* Mrs Beeton's *Household Management* leaves us in no doubt about this issue: 'Barndoor fowls are less fat than, but far superior in flavour to, the fowls fed in close coops for the town market.' Her recipe for roast chicken requires the addition of basting fat – hardly needed in roasting today's broiler. Interestingly, despite her frequent use of cream in recipes, she insists that after roasting, the fat should be discarded – 'drain off every particle of fat'.

close to 8 per cent and by the early 1980s it had risen to 22 per cent. As a proportion of the calories, the carcass fat of the modern broiler chicken had, at the time of writing, exceeded the proportion as protein. The drive for 'weight gain' resulted in fat gain. Nutrient gain was not part of the design.

Perhaps readers might be interested to see the extent of the discrepancy in modern broiler chicken production which is going the same way as beef. Using official data, we have compared it with partridge which is wild.

Table 7 Proportions of protein and fat in domestic and wild fowl

	protein g/kg	*fat* g/kg	calories *protein*	calories *fat*	*energy ratio fat:protein*
broiler chicken					
males (56 days)	97	211	388	1899	4.89:1
females (56 days)	94.3	233	377	2097	5.56:1
wild partridge	120	30	479	270	0.56:1

Protein is taken as a reference point because most would assume that animal production is for protein. However the protein-rich part of the animal also contains iron, trace elements, vitamins and essential fatty acids. The data in the tables illustrate that the ratio between fat and protein is in the order of a tenfold difference if man-made animals are compared with their wild counterparts and hence their starting points.

Unfortunately the fats that are bad for our health were, at the outset, good news for the food industry. Because polyunsaturated fats go rancid faster than saturated fats, the soft fat in pigs was a serious problem but was overcome by feeding carbohydrate-rich high-energy feeds. The hard saturated fat produced by such high-energy feeding keeps much better: try keeping a jar of fish oil for as long as you can keep a tallow candle!

It would be wrong to use this argument as an attack on the food industry. When one considers that the people of London or New York need to consume 35 billion calories of food per day and the food industry has to prepare and ship 5 million metric tonnes of food a day into these dense conurbations, one realises just what a prodigious task they perform. It is little wonder they were interested in shelf-life. When the food industry began to be faced with such logistical problems, it had no guidance from medicine or science. It had to feel its way forward. Its difficulty now lies in sorting out the scientific evidence, the increasingly higher level of public awareness and then deciding what to do about it.

With the industrial revolution came many benefits and many drawbacks. Butter had been a food for the rich. Chemists found that they could make butter out of surplus beef fat (tallow), or at least a substitute for it. The poorer people were then able to put a butter substitute which was called margarine on their bread. However, there was not enough to satisfy. This innovation led to the use of hydrogenation techniques which took polyunsaturated oils from fish and whales and made them saturated. After a bit of deodorising they began to claim that 'you could not tell the difference'. Indeed any oil could eventually be hydrogenated and made like butter.

There were two effects. First, the addition of hard, butter-like margarines for putting on bread and for cooking with, meant adding another saturated fat load to the diet which had not been there before. Secondly, the increase was achieved by destroying the original essential polyunsaturated content of the food fat. So the saturated fats were increased and the essential fats decreased.

More recently, the evidence showing that saturated fats cause atherosclerosis and thrombosis, whilst essential polyunsaturated fats are needed for the health of the blood vessels, has persuaded some, such as Unilever (who were the first commercial organisation to recognise the evidence), to produce margarines which were made with seed oils rich in the essential fatty acid, linoleic acid. They and others also made available vegetable oils for salads and cooking in their natural form, as oils low in saturated fats and rich in essential polyunsaturated fats.

In the USA the use of soya bean oil, which contains both linoleic and alpha-linolenic acids, expanded during the late 1960s and early 1970s. In Australia, a traditional cattle country, 60 per cent of the butter had been replaced by the polyunsaturated margarines by 1981. Also in Australia, and indeed the USA, meat producers have recently developed aggressive programmes to eliminate the fat animal. (Norway has moved into salmon farming.) The key message to farmers at the New Zealand Ministry of Agriculture's open day at Ruakura in 1985 was 'Fat is not for sale'. Despite all this, the beef and poultry industries have resisted the trend and moved but little towards real meat which should be a rich source of essential polyunsaturated fatty acids and many other nutrients.

Similar changes occurred in the use of carbohydrate sources. Sugar was extracted, purified and concentrated from an overgrown grass. Wheat was developed from a sort of grass seed but selection and processing concentrated on the carbohydrate-rich white part. The food manufacturers wanted the endosperm (the white inside), not the nutrient-dense sperm nor the fibre-rich coating. These were thrown away or fed to animals. As it happens the germ is a concentrated packet of nutrients. It contains high concentrations of E and B vitamins, trace elements and essential fatty acid rich oils.

Bread is conceptually a seed food, so removal of the germ to make white bread, or selecting for a large endosperm, dramatically changed the basic nutrient balance. What is amusing is that the recent developments in food technology quite accidentally turn out to be a method of putting at least the oils back. Conceptually, the 'Flora' approach to polyunsaturated margarines for spreading on bread, is restoring the essential fatty acids and vitamin E which would have been in the seed germ in the first place! Also the work of people like John Yudkin, Commander Cleave and Dennis Burkitt has led the manufacturers to start putting back the fibre into food and finding that they can profit as well.

Likewise, fruits were originally developed for their fluid content and became a good source of vitamin C. However, the idea of the oil-rich seeds which they may have contained was lost. Some oil-rich seeds were developed such as walnuts, almonds and hazelnuts, but it is an extraordinary coincidence that recent developments in the use of both animals and plants have diminished the oil or essential fatty acid component, whilst at the same time increasing the energy supply in the form of saturated fats, sugar or purified carbohydrates. Whilst this switch in the balance can be most clearly identified and measured with regard to the relative loss of essential fats, it has also taken away other important nutrients including trace elements, vitamin E and fibre. In developing countries, where they became dependent on rice, much the same happened. The rice was polished in an analogous manner to the de-germing of the wheat.

These changes in diet, especially the exchange of essential fats for saturated fats, are precisely those changes which in experimental animals have been shown to induce atherosclerosis and thrombosis. Seeing the effect of those changes in experimental animals of every sort, means that you would have to come up with a pretty good argument to say that the same changes in human diet would not produce the same changes in the human species. No such argument exists, although some try to escape by assuming that man will adapt. Unfortunately they say this without realising that only five generations have been exposed to this change which coincided with the industrial revolution. Once that is explained the argument collapses.

The changes which have occurred in food composition are very recent but were brought about when neither medicine nor science were able to act as a guide. We learnt about scurvy, pellagra, vitamin A blindness, about vitamin D and rickets and all the rest through our mistakes at the beginning of this century. In effect, we lurched from one mistake to the next. These were short term problems. We solved these and we now have to solve the long term problems like heart disease, cancer, multiple sclerosis and arthritis.

We can identify modern intensive farming and its technological mimics of high saturated fat foods as a mistake like the purification of carbohydrates. Both focused on producing the energy rich, not nutrient dense foods. Independently, they were harmless: butter is not poison and sugar is sweet – a little of either is of no consequence. But put them all together in large amounts and you quickly reach a breaking point.

From a biochemical viewpoint, the changes in terms of fats alone mean a large increase in the intake of non-essential fats and a reduction in the essential fats. Excessive sugars and carbohydrates can only be converted and stored in the body as non-essential fats. It happens that science is now aware that the essential fats are needed for growth, development and health of the brain, the vascular system and the immune defence mechanism. Swamp the essential fats by non-essential fats and then reduce their amounts and you would expect vascular, nervous and immune system disorders. From the present scientific evidence there is a strong probability that this is what we are seeing: a high death rate from heart disease, stroke, multiple sclerosis, breast and colon cancers which occur with a uniquely high incidence only in those countries which have developed food in this way.

The Joint Expert Committee on Human Nutrition called by FAO

and WHO in 1977 expressed concern about this system of modern animal agriculture and made specific recommendations for correction. More recently a NATO workshop held in Selvino, Italy, in 1986 took the argument a step further. Its recommendations included the following:

> In view of the very serious nature of nutrition related disease in certain technically advanced countries, it would be undesirable for those countries to transfer their principles of food technology and agriculture to other countries in which cardiovascular disease, breast or colon cancer are not at this time a problem.

Biologically speaking, man is still a wild animal and there is no reason to suppose that his biology is adapted to anything other than wild foods. There simply has not been time for any selective evolution to have changed mankind as mankind has changed its pattern of eating. Man has no more developed an adaptive defence against vitamin A blindness in India or volvulus in Africa than he has against heart disease in Northern Europe and America. We shall see in the next chapters further evidence of the way the changing diet of *Homo sapiens* is affecting his physical nature, shape and form.

13 The scientific consensus on heart disease

Homo sapiens evolved on the wild, unsophisticated foods of nature for 99.8 per cent of his existence. He was astonishingly successful. His population grew and eventually his need for more and more space and food saw to the demise of other species and their habitats. There are now about six billion of us on the planet and but a few hundred chimpanzees or gorillas. On the road to this total dominance, the balance between population and natural food supply began to make life less comfortable. Donald Gebbie describes how the Australian Aborigines and the Polynesians had an answer to population and food supply control for 50,000 years until Captain Cook arrived.

The rest came to depend more and more on domesticated plants and animals. In different parts of the world, techniques of agriculture and climates were different and so the food they produced was different. The drift away from basic evolutionary food composition occurred at different times. Man progressively took the place of nature and it was he who decided what he wanted in a plant or animal food. Some developments (like winter greens) were hugely beneficial but others spelt disaster.

It is easy to see how certain directions in agriculture and food production led to the nutrition-related disorders of India and Africa, how deforestation produced dust bowls, but not quite so easy to recognise that we in the West made similar mistakes.

It is now accepted by the Food and Agriculture Organization and the World Health Organization in their reports of 1978 and 1982 and by the Surgeon General's Report on Health in the USA, mid 1988, that there is a nutritional link between diet and heart disease. There is also evidence and concern that multiple sclerosis, arthritis, breast and colon cancer fall into the same category of nutrition-related disease.

These diseases are common only to the technically advanced countries, with Japan and the Mediterranean countries being important exceptions. More disturbing is that when parts of Third World countries become affluent, they copy the West and, for the first time, we hear of heart disease and the Western cancers in those who adopt its way of life.

A hundred years ago heart disease was something of a rarity, but in Europe and the USA today about one man in four will have a heart attack or a stroke before he reaches retirement age.

Initially there was little public consciousness of the alarming increase in heart disease: the problem crept up on people, partly perhaps because doctors did not like to worry and depress their patients. Worry, after all, could make it worse. But in the early 1970s, in one small country, this protective cocoon broke down.

Statistics collected by the World Health Organisation on heart disease in different countries showed that Finland had a much higher death rate from that cause than any other country in the world. It was so bad that it was commonplace for a Finnish mother to grieve the loss of her son in middle life from a heart attack. It happened that this was a time when Finland was opening up communications and Finns were travelling to different parts of the world; those who returned from visits abroad, either to the eastern bloc or to the west, brought news that in other countries heart disease was not so devastatingly frequent. As this realisation took hold there was a general demand by the people that the government should do something about it.

The result was the North Karelia project. In proper scientific fashion the organisers, directed by Dr Pekka Puska, took two sections of Finland – North and South Karelia – and put the programme into action in North Karelia only, leaving the other half as a control group. This would allow a comparison to be made and the effect of the programme to be gauged. In North Karelia a massive campaign was launched which ranged from educating people on the effects of smoking, exercise and nutrition, right through to changing the foods available in shops and supermarkets. Even the sausage was to be different. And it worked. Seven years later mortality from heart attacks was down by 25 per cent and from strokes 30 per cent.

But was the experiment a success? What was happening meanwhile in South Karelia? The people there had been watching what was happening in the north, and had begun, of their own accord, to change their habits too. Heart disease went down in South Karelia, although not to the same extent as in the north. Now this could have been taken as reasonable proof that something could be done to reduce heart disease, but many objectors saw it differently. To them it meant that the reduction in North Karelia was not due to the programme at all: since the rate was falling in the control group as well as in the experimental area there must be some different cause altogether. The Finns themselves believe it was no accident. It was just impossible to keep the information from the control group.

Unfortunately, Finland is not blessed with warm sunshine, fresh fruits and green vegetables or sea foods typical of the diet in the South of France. Being ice-locked, their choice in food development was to use stored feed for cows and they had ended up with the highest butter fat consumption in the world. This difficulty makes their achievements all the more remarkable. How far they will be able to continue their success in the face of this difficulty, and how they do it, will be most interesting.

There was, however, rather a good control group for the Finnish experiment in the country from which a lot of objections came – Great Britain. Those who sold saturated fats were prominent among the objectors and, probably as a result of their lobbying in high places, practically nothing had been done in Britain to prevent heart disease. There was talk but little action. In 1974 the Department of Health published a report on diet and the prevention of heart disease which recommended cutting down the intake of fat, particularly saturated fat, and in 1976 the Royal College of Physicians produced a report under the chairmanship of Professor Shaper, recommending more detailed action. Two members of the Royal College's working party, Dr Keith Ball and Dr Richard Turner, set up the 'Coronary Prevention Group'; but they had to get by on an annual budget of less than £30,000 at a time when the manufacturers of questionable foods were spending hundreds of millions of pounds a year to press their produce on the consumer.

So, while the Finns benefited, the most badly affected part of Great Britain – the North, Scotland and Northern Ireland – swapped places with Finland in the death league and came first. The irony was that European scientists were welcoming the inaction in the UK, saying, 'we need one country as a control'! The reality is that in countries where the government has taken direct action, or where there has been strong public awareness of the issues, as for example in the USA*, Canada and Australia, there has been a drop in the levels of blood cholesterol and a fall in the death rate from heart disease.

There are still those who argue against the role of diet in heart disease. They usually adopt five tactics. The first is that there is no argument; the second is to say nothing about the historical changes in diet; the third is to claim there is no proof; the fourth is to ignore the scientific evidence from experiments with animals; the fifth is to insist that man would adapt to any changes in diet. They argue that nothing should be done about diet. They usually forget that Darwin said adaptation and natural selection worked hand in hand with 'conditions'.

As for disagreement: of course those at the cutting edge of the science have their hobby horses and rightfully are often the most argumentative; they are prima donnas and not unnaturally would like their own versions of the story to win the prizes. There is, however, a core level on agreement that is solid. The Food and Agriculture Organization of the United Nations seldom calls a conjoint meeting with the World Health Organization and when it does, the expense entailed ensures that the position has been reached where the experts will actually be able to say something useful. When the FAO/WHO Expert Consultation on Dietary Fats was called in 1977 there were already 18 similar national and international recommendations for action on diet and heart disease.

In 1982 the World Health Organisation called again on international opinion to write a further report. The original FAO/WHO committee was comprised mainly of scientists, and dealt with the global issue. The later WHO committee was specifically medical and addressed the narrower problem of heart disease. They both said much the same. By the time the matter had twice hit the international level, it was becoming apparent that the UK was being left out. A previous report by its Department of Health in 1974 and the Royal College's efforts of 1976 had largely been

* At the beginning of the 60s, the image of the American businessman was a corpulent fellow with a cigar permanently in his mouth. Today he is a lean, crew-cut character, who knows about cholesterol, has running shoes in his brief case and believes that 'Lunch is for wimps'. The seeds for this revolution were being sown in the early 60s when books such as Wheeler's *Fat Boy Goes Polyunsaturated* (Doubleday, 1963) and Blumenfeld's *Heart Attack – Are You a Candidate?* (Eriksson, 1964) were bestsellers in the USA. In Australia, the Heart Foundation has, since its inception, been aggressively concerned with prevention and enjoyed co-operation from the meat industry as well as margarine manufacturers and many other sectors of the food and catering industry.

ignored. So the Department of Health formed a National Advisory Committee on Nutrition Education (NACNE), which under the chairmanship of Phillip James wrote a discussion document on nutritional guidelines for health education in Britain. The report was promptly disowned: but too many people knew about it and it was published under the aegis of the Health Eduction Council in 1983.

A Joint Advisory Committee on Nutrition Education was then set up to explore the practical implications. Interestingly enough, the agricultural and food industry implications had already been discussed in detail by a succession of meetings and reports from the Coronary Prevention Group and the Centre for Agricultural Strategy in the 1970s and early 1980s and both had also been ignored. So it is hardly surprising to note that mortality from heart disease continued to climb in the UK during the late 1970s and early 1980s whilst in the USA, Canada, Australia and Finland the death rate was falling.

In the meantime the wider health issues had been faced in 1976 by a US Senate Select Committee on Nutrition and Human Needs under the chairmanship of Senator George McGovern. It identified major public health issues in the United States, such as heart disease, cancer, hypertension, obesity and dental caries as being linked to inadequacies of the American diet and, thus, preventable. All this was to be confirmed by the 1988 USA Surgeon General's report.

Sir Richard Doll's 1968 thesis on contrasting cancer incidence in different countries led to a report in the early 1980s, by the National Research Council of the USA Committee on Diet, Nutrition and Cancer. Why is it that Americans and Northern Europeans have a very high death rate from breast cancer whereas Japan, Thailand and other countries where fat consumption is low do not?

In essence, all ask for a reduction in the total amount of fat and particularly in the saturated fat content of the diet. The conjoint FAO/WHO committee had identified the need for the essential polyunsaturated fats and even made recommendations for corrective measures with regard to animal and plant production. In 1986 a NATO meeting was held in Selvino, Italy. Organised by Professor Claudio Galli from Milan, its aim was to discuss the relevance of technologies to food and health. With a high representative of food technologists, some thought there would be dissent, but in fact they added a seal of approval.

An International Consensus meeting on what the doctors should do was then held in the USA in 1987. There was much squabbling but a consensus did emerge: we should know whose blood cholesterol is above 240mg per 100ml and do something about it. Of course there are different views about details but that is to be expected of any subject. In all some 65 different committees have discussed these issues and yet this tower of Babel has actually reached a remarkable consensus. Admittedly, all may be out of step except 'Our Johnny' but it is unlikely. We may not know all the answers but we obviously know enough to at least experience a fall in mortality where action is taken: so much for disagreement.

As for the laboratory evidence: the library shelves groan with data showing that feeding saturated fats increases blood cholesterol, damages the arteries, stimulates the blood clotting mechanism and, indeed, affects blood pressure and the immune system. By contrast, the essential polyunsaturated fatty acids are needed for reproduction, brain growth, vascular system development, cholesterol excretion, control of blood lipids, blood pressure and other important regulatory functions.

Quite simply, too much saturated fat, sugar and refined carbohydrates in the diet compete with and interfere with the use of these essential polyunsaturated fats and increase blood cholesterol levels which, simply on its own, leads to hardening of the arteries. Put competition and hardening together with leaky cell membranes, raised blood pressure and an increased tendency to thrombosis and the ground is prepared for coronary thrombosis or stroke. Hence the mechanism leading to the disease of Western countries can be spelt out in some detail.

That the saturated fats played little part in our evolution has already been discussed. So if diet did change, has man adapted to the change? Here nutritionists and geneticists are in equal agreement. They met in 1984 to discuss *Nutritional Adaptation in Man* (Blaxter & Waterlow, 1985). They concluded that there was little evidence of human adaptation to over- or undernutrition. Hence the contrasting disease patterns of India, Africa and Europe reflect exactly the failure to adapt to the contrasting changes in food over the last few centuries. Alternatively, one could say that the diseases themselves are the adaptations. Certainly science historians would see them as examples of non-adaptive evolution: a subject on which Darwin had uncharacteristically little to say.

One piece of evidence on adaptation came from an interest in the reason for the small stature of people in Third World countries. Because they were consistently undernourished there was, it was thought, a selection pressure for smaller people. This idea was tested by studying periods of food shortage and famine. Famine should weed out the children with the genes for larger bodies. Selection pressure would favour those with genes for small body size and so gain the advantage of a smaller requirement for food. The studies showed the opposite: during a famine, it was not the smaller children that survived but the larger children. Presumably the larger children had been better fed in the past.

If food has its dangers it can also come to our rescue. There is evidence to suggest that certain types of food can give protection against cancer, including cancers in parts of the body that have no direct connection with the digestive system. The Japanese, for example, smoke more cigarettes than anyone else, over 6000 a year per head compared with about 4200 in Britain and the USA, yet their level of lung cancer is only about one fifth of that of the USA.

They also have very low rates of colon and breast cancer. By contrast, they have a very high rate of stomach cancer although this mainly affects the poorer sections of the Eastern populations and may also be related to food. Despite the incredible overcrowding in Japan, the fast pace of life and intensive industrial activity, the Japanese are also protected from heart disease. However, the children of Japanese who migrated to America at the turn of the century, who have been brought up in the American style of life, now have the same incidence of heart disease, colon, breast and lung cancer as their American hosts, so it is clearly not a genetic protection that explains the low incidence in Japan.

This is not to say that genetics are irrelevant. One person will be more susceptible than another for reasons of genetic heredity. However, the nutritional change of the migrants has overwhelmed any general genetic difference that might be thought to exist between Japanese and Americans: it has swamped genetic protection and unleashed susceptibility. It is a matter of gene expression responding to a new nutritional environment.

Dr Hugh Sinclair of Oxford was really the first to pinpoint the positive contribution of seafoods in the diets of the Japanese, Mediterranean countries and the Eskimoes is important in the prevention of heart disease.

It also clear that the nature of the fat is important. Ancel Keys' seven countries study revealed the link between saturated fat and heart disease because the study included the Mediterranean countries whose fat intake was equal to that of Northern Europeans or Americans. Their fat, however, was different, being based on olive oil which has a very low proportion of saturated fat and is mainly a monounsaturated fatty acid, oleic acid. Olive oil also provides an adequate amount of linoleic acid and this, combined with the Omega-3 fatty acids from the Mediterranean sea foods, explains the low incidence of saturated-fat related disease. Similarly, Eskimoes have a high intake of fat but it is polyunsaturated. The blubber and fish meat contains a wealth of Omega-3 fatty acids. However, they do eat a lot of seal meat, which like wild buffalo meat, contains a relatively low proportion of fat, but what it does contain is structural and has a high proportion of both Omega-6 and Omega-3 fatty acids. It is also worth mentioning that with the essential fatty acid rich foods goes a range of accessory nutrients, anti-oxidants and trace elements as would be expected.

Nor do the Japanese die from something else instead. They actually live significantly longer than any other large nation. The traditional Japanese diet contains a lot of seafood and fish, often eaten raw, but little from land animals. There is a memorial, built in 1930 in the Gyokusenji Temple, Shimoda, to mark the spot where the first cow was killed in Japan for human consumption. The use of the cow in this context was of course prompted by Western example. Curiously, the chains around the memorial displayed a notice which say they were a gift of the US Navy. The significance of the gift of chains is somewhat obscure. Maybe someone foresaw that this event represented the first step in chaining Japan to American food and disease!

The disease of the arteries which leads to a heart attack is not just something which appears in middle life. It is a lifelong process. The risk of death from heart disease is several times greater in the lower socio-economic groups. Professor Barker and Dr Osmond of the Medical Research Council's Epidemiology group in Southampton studied infant mortality of 50 years ago in 212 different regions in the UK, reflecting different socio-economic groups. Rich and poor people were compared: historical data of current death from heart disease was compared with past infant mortality as a measure of poverty. They concluded that, 'Pre- and early post-natal influences, including infant nutrition, are determinants of the risk of ischaemic heart disease and stroke.' (i.e. early nutritional influences predispose to heart disease in later adult life.) (Barker & Osmond, 1986)

In the 1960s, Dr G. Osborne, an Australian pathologist, studied the arteries in the hearts of 1600 babies who had died accidentally or from problems not related to the heart. He found early damage already in the blood vessels. He tried to work out why some were damaged and some were not. The only conclusion he could draw was that the lesions were seven times more frequent in bottle-fed as opposed to breastfed babies (Osborne, 1967). This is only surprising if you do not know that human milk is designed for the growth of the brain and the vascular system whereas cow's milk is for muscles and skeletal growth. The highly complex vascular system that enables our hands to play the piano or guitar, is a hoof in the cow.

Further evidence came from postmortems in the Korean war. Young American soldiers of 18 to 20 years of age, who had been killed, were found to have atherosclerosis, the blood vessel disease that leads to heart disease. Danish research workers have found that the umbilical arteries of babies born to mothers who smoke have atherosclerosis!

So, what happens to blood cholesterol and blood pressure, the two most powerful predictors of death from heart attack in children? Africans and other people in whom heart attacks do not occur, have low blood cholesterol levels and their blood pressure does not rise much or at all with age.

Crawford & Hansen (1970) asked if European children and African children were born with different blood cholesterol levels or the same? They turned out to be the same in the first year of life but by six to eight years of age, the blood cholesterol levels of the European children were already raised in comparison with their counterparts from low risk communities. Several laboratories have reported the same findings. Drs Lauer, Clarke and Rames of the Iowa division of paediatric cardiology reported that blood pressure also starts to rise in the same age group. Babies and young children across the world get much the same exercise, do much the same things and are unlikely to be smoking. So the explanation must relate to food.

For these two predictors of atherosclerosis to be so identifiably different at this early age it must be that the process has been in operation well before the ages of six to eight to force its way through. Indeed, the evidence on heart disease is serious enough on its own. However, the fact that one can pick out children from high risk groups simply by measuring their blood cholesterol levels at six years of age, poses a very serious question. Bearing in mind the relationship between the nervous and vascular system, does vascular disease affect foetal and hence brain development? If so what impact will this have on future generations?

The highest risk group within the UK for heart disease is the lowest socio-economic group and it is the same low socio-economic group that is at high risk of low birth weight and associated handicaps. Of the non-genetic handicaps, educational subnormality is the commonest and is strikingly more common in the lower socio-economic group. Much of this kind of data is described in an excellent book by Herbert Birch and Joan Gussow entitled *Disadvantaged Children, Health, Nutrition and School Failure* and in *Malnutrition and Brain Development* by Myron Winick (Birch & Gussow, 1970; Winick, 1983).

The contrasts in subsequent IQ in relation to weight and head circumference at birth across the socio-economic scales must be due to parameters set before birth. Data from developing countries is what people mostly talk about rather than the data on our own children. However, the one situation supports the other. The term 'the cycle of deprivation' was coined to describe how maternal IQs correlate with maternal and infant nutrition. Prescott and others in a scientific collection of papers entitled *Brain Function and Malnutrition* say that there seems to be strong evidence that 'Malnutrition *per se* not only affects the expression of genetic potential in physical development, but acts upon intellectual development . . . Those affected are limited, making it difficult for them to be wholly incorporated in the socio-economic development of the country (Prescott, 1975).'

How is it possible, the reader might ask, that foetal growth retardation and heart disease have any common ground? Wendy Doyle's studies of maternal nutrition during pregnancy in contrasting socio-economic groups may provide an answer. Birth weight, head circumference and placental weight go together. The nourishment for foetal growth has to come from the placenta which has to grow first to do its job. Heart disease is basically a disease of the blood vessels feeding the heart muscle. Clog one up and the heart muscle gets no nourishment, heartbeat control becomes chaotic or suddenly stops altogether. The placenta is basically a rapidly growing, brand new system of blood vessels. Hence nutritional principles which produce diseased blood vessels and finally heart disease may do the same when the placenta is growing.

This is precisely what the study by Wendy Doyle, Dr Kate Costello and their colleagues is suggesting. Normally, women are better protected from heart disease than men by virtue of their hormones and higher metabolic efficiency for fats. Once past the menopause that protection is lost. However, the protection is not total and as the placental vascular system grows so rapidly, it would not be surprising if this was the vulnerable point.

The placentae of foetal-growth-retarded babies are not only small but usually show evidence of blood clots. Myron Winick of New York has independently found that poor placental development associated with multiple thrombi may be an important cause of foetal growth retardation; the findings in London by Wendy Doyle and others would be consistent. The common denominator of the highest risk of foetal growth retardation arising in the same low socio-economic group which has the highest risk of cardio-vascular disease may well be in the common vascular considerations of the blood vessels of the heart and the placenta, on which both depend for their function.

There is one further doubt that is worth discussing. The doubters still want clinical proof, but they may not be asking a sensible question. When it is realised that people are addressing a lifelong ecological problem, then the failure to obtain dramatic results with cosmetic dietary alterations at the age of 50 is to be expected. One should not expect too much until preventive measures are taken by whole families.

From a philosophical point of view proof of anything is impossible. Even physicists, who are the closest in science to the nature of the universe, find the concept difficult. Probability yes, but proof is more difficult. Various trials have been done in an attempt to change 50-year-olds' diets in some way and lessen their risk. The doubters say the diets did not work or they proved that they may have been saved from heart disease but died of cancer instead.

For example, the multi-million dollar trial dubbed MR FIT in the USA didn't work. People then said that must mean the whole theory was wrong. In fact, they were dazzled by the cost of the trial. At the end of the day, there was little more than a 2 per cent difference in blood cholesterol levels between the test and trial groups. Actually the conclusion from MR FIT is that it confirmed that with no reduction in blood cholesterol you cannot expect a reduction in mortality. They had not even reached the starting gates!

If the results of all the trials done are examined critically, they actually do provide positive evidence. Professor Richard Peto, from the epidemiology unit in Oxford has pointed out that the measure of success of the execution of a trial based on the lipid hypothesis would depend on whether or not blood cholesterol was lowered. The measure of the success of the test of the hypothesis is that if blood cholesterol is lowered, mortality from heart attack should also be lowered. When the statistical analysis was done on all the trials, the extent of cholesterol lowering was the measure of reduction in coronary incidence. The Oslo trial was the best example with a reduction of 49 per cent in mortality in a high risk group over seven years. It was the trial which did the most in terms of diet, blood lipids and reduction of mortality.

Dr Ancel Keys and his colleagues had concluded from the seven countries study in the early 1950s that death from heart disease correlated strongly with the amount of saturated fat eaten and with the amount of cholesterol in the blood (Keys, 1980).

Today the story is substantially the same except for one important new development. We now know the other side of the equation: we now know that the body and the arteries require a special type of fat, the structural lipid, and that these are the essential polyunsaturated fats. Dr Hugh Sinclair was again the first to point this out in 1956. Even though the data was sketchy at that time, Dr Keys worked out a mathematical formula to predict the degree by which blood cholesterol will fall if you change from a saturated fat to polyunsaturated fat diet; and it works!

Further evidence for the protective role of the essential fatty acids came from a study by Dr Michael Oliver and his colleagues. Scotland now has the highest mortality rate from coronary heart disease in the Western world. Sweden is middle of the road. The comparison of the death rates in Edinburgh and Stockholm showed that the Swedes with a lower death rate had the same blood cholesterol levels so what could it be that protected them? By studying the composition of the storage fat in the two groups of people, Dr Michael Oliver found that the Swedes had higher levels of essential polyunsaturated fatty acids, particularly linoleic acid, in their body fat.

Experimenters used polyunsaturated fats as the *control* to study the effect of feeding saturated fats on arterial disease. More recently, it has become apparent that to build healthy arteries and maintain good blood flow, it is important to have an optimum balance of the Omega-6 and Omega-3 families of the essential polyunsaturated fatty acids. That optimum balance and its Omega-6 and Omega-3 essential fatty acids as they occurred in natural foods is almost certainly that which was achieved throughout the period of human evolution.*

Europe has lurched from one nutritional problem to another: scurvy and rickets are understood because of past mistakes. Meanwhile, medical science in Asia is still learning from the errors of recent history. One thing is agreed: the present incidence of ill-health in East or West is unacceptable and the task set by WHO of Health for all by the year 2000 is an attempt to escape from past errors and continue the forward thrust of human evolution. The difficulty, as always, is the courage to act on the evidence. But no action is a cop-out: it is tantamount to action because lifestyle and food are being changed all the time – willy-nilly.

Looking at the scene of human nutrition from this perspective it is obvious that throughout human evolution, man relied on wild foods. His physiology was initially adapted by, and is still adapted to, wild not modern foods. He spent 99.8 per cent of his existence as a species living on wild foods. In Northern Europe and America, as little as 0.006 per cent of his time has been spent on the modern foods created by the enclosure of animals and even less since the industrial revolution. In Western countries we have developed animal proteins but the wrong kind of fats and neglected those that are needed for the nervous and vascular systems: indeed, those needed for the highest specialisations of *Homo sapiens*. That is what is worrying: the high specialisations which led to the success of *Homo sapiens* as a species are now specifically under attack. And

* Glaxo, one of the world's leading pharmaceutical companies, developed in 1987 a special concentrate of the Omega-3 fatty acids from fish to be made available through the National Health Service in the UK because of its properties of lowering blood lipids and reducing the risk of thrombosis. Are they simply trying to restore the balance which occurred during our evolution?

they are under attack because we have changed the food which underpinned their evolution.

The different diseases set against different nutritional backgrounds of the different countries tell us first, that nutrition has a profound effect on human physiology and performance. Secondly, they tell us that the change in food composition has occurred at a rate too fast for any selective response. Once again we can link the omission from evolution theory of Darwin's 'conditions' to the incremental growth of degenerative diseases. The response to the rapid change in food structure is being expressed in the form of atherosclerosis.

We do not know all the answers to heart disease nor to the problems of the East and nor is it all just a question of fats. But there is enough simple evidence staring us in the face should we choose to look. Given the will and the application of intelligence, by the middle of the next century, people will look back on the ravages of vitamin A blindness and endemic heart disease, cancer, multiple sclerosis and arthritis of today, in the way we now look back on the ravages of rickets of the 1920s.

14 Man, evolution and the future

In this book we are not suggesting that natural selection is contradicted by the theory of substrate-driven change: it is not an 'either/or' situation. We are talking about two powerful mechanisms which both operated in the origin and shaping of species. From the evidence we have presented, it is difficult to regard substrate-driven change or plastic heredity as a trivial force as it seems to have been behind many of the great evolutionary thrusts, including the origin of life, the nature of the first living forms, the emergence of oxygen breathing animals and the shaping of our own species.

An important distinction between selection and substrate-driven mechanisms, is that the latter offers a powerful predictive potential which, on its own, natural selection lacks.

At the present time, one in three or four men in the USA or Northern Europe will have a heart attack or a stroke before retiring from work. A large proportion of those with heart attacks will be dead within a year; those with strokes will probably remain paralysed or partially so for the rest of their life and are likely to have another attack. These disorders have their fundamental cause in the vascular system. At its best it is simply stress and we need only to relax. At its worst, it is the tip of an iceberg.

The fact that cardiovascular disease specifically hits one section of the human population means, quite simply, that one of the key specialisations of *Homo sapiens*, the vascular system, is under attack.

It is possible to interpret this contrasting incidence of nutrition-specific, degenerative disease as substrate-driven evolution in progress. There is evidence of abnormal distributions of different types of high and low density lipoproteins in the bloods of those with atherosclerosis and ischaemic heart disease. It is quite plausible that these are a reflection of gene expression being exposed by food.

Atherosclerosis undoubtedly affects the form and function of the vascular system and ultimately the function of the individual.

If that is so, then the evidence discussed in this book on the evolutionary link between the nervous and vascular system raises the speculation that the brain and the nervous system will be the next to go, unless, that is, something is done.

The brain is better protected because it is built in the womb of the mother and would, predictably, experience a period of grace – a time lag. But there are those who consider that the brain is already showing the signs. Multiple sclerosis is the commonest disease of the brain and nervous system; it attacks between the ages of 15 and 40, follows an unpredictable course with some sufferers experiencing no more than a few bouts, whilst in others each successive attack sees further malfunction or paralysis depending on which part of the brain breaks down.

Professor Roy Swank of Oregon University, USA, has described the distribution of multiple sclerosis from country to country as roughly following that of heart disease. Following this lead, Professor Swank's studies have shown how dietary management can help. Such evidence has led to three double blind trials offering a chink in the armour of this disease.

Views differ as to the cause but the possibility that a nutritional/genetic predisposition is set in place during brain development, has led Action Research on Multiple Sclerosis in London to develop a programme of nutrition coupled with physiotherapy, psychological counselling and neurological monitoring. Although there is still no cure for the disease, their data provides further evidence (as in the case of heart disease) in favour of nutrition intervention, and opens up a new research direction. Again, like heart disease, the real difficulty is in challenging in late life a process which has its origin perhaps as early as foetal and infant development.

Alzheimer's disease has suddenly become an important issue. It is a process of rapidly accelerated ageing or, rather, a pathological shrinking of the brain, leading to senility. The reason cannot be ascribed to living longer as people are affected in their 40s and 50s. Although speculations and ideas abound, no cause or cure is known for Alzheimer's disease.

Furthermore, despite all the advances in medical science from antibiotics to cardiac surgery, artificial arteries, valves, bypasses and transplants, present life expectancy in the West beyond the age of 40 is no more than two or three years better than it was at the turn of the century. Because so many children died then in the first few years of life, figures for the average lifespan gave a false impression. This is not to say that the threescore years and ten of the Bible is the limit. Some say that, whilst medical science has been making great strides, something else has been working in the opposite direction. The important point is that we should be able to reach old age (whatever that is) in a fit state and not as vegetables.

Others are worried that if we cure heart disease we will increase the risk of cancer. In 1985 the Swedish Society for Medical Science and the Swedish Nutrition Society jointly called an international conference to discuss the question: 'Are two different diets needs for the prevention of cancer and heart disease?' The conclusion was that we do not yet have enough evidence on cancer but, so far, the data indicates that the action being taken to prevent heart disease will be similarly beneficial in cancer prevention.

If we solve the equation between diseases of this sort and food, then we will not only eliminate much distress but people will look forward to an expanded lifespan in full health. The present fear of losing our intellectual and physical faculties will be replaced by the expectation of richness derived from the true pleasure of living with health and the time to enjoy it. The idea that if you remove one disease it will be replaced by another is not supported by the evidence and is defeatist. Instead of capitulating to the forces of nature, we need to answer the question, 'What are the optimum conditions required to continue the advance of the human species?'

In this book we are presenting the case that nutrition is not just concerned with ill or good health but that nutritional chemistry was, and is still, a fundamental evolutionary force. Food is of such a commonplace nature that it is taken for granted and its qualitative relationship with long term biological considerations is overlooked. The historical change in disease patterns, the contrast in incidence from country to country and, importantly, the socio-economic contrasts within a country, suggest we are witnessing a signal of the potential power of food as a dominant factor in evolution.

The value of introducing nutritional chemistry as the major directive force in evolution is that it offers a predictive power lacking in the theory of natural selection. Nutrition or substrate theory accords with the view of the recent change in human body size and incidence of degenerative disease as evidence of human evolution still in progress. This conclusion logically leads to predictions about the way in which the nature and form of the human species might change. The change in body size since the turn of the century, and in disease patterns, indicates that these substrate-driven changes occur at a fast speed which outstrips selection processes. Hence nutrition theory poses serious questions as to the impact of present day food and agricultural policies on immediate and future generations.

Some people argue that it is wrong to change our food but they forget two facts. First, we have already changed our foods in several major dimensions. Secondly, a policy of no change is unworkable because people, industry and agriculture are changing things all the time. The point is that errors have already been made and it is not a question of changing foods, it is a matter of *correcting* the errors and moving forward from there on the basis of science rather than folklore. We now have much of the knowledge to guide us into the future: the stakes are too high to gamble.

The reader needs little reminding of the reasons why this issue should be addressed. Today, millions face famine and the lives of many more are dominated by food shortages whilst two billion people are without clean water. It is estimated that 450 million people suffer from hunger or malnutrition in developing countries and UNICEF expects 14 million children to die before their first birthday in 1989. Population Concern, a body of people headed by scientists, points out that more people have died in the last 100 years from hunger and malnutrition than in all the wars, murders and accidents on which we have recorded data. There are 42 million blind people in the world, 80 per cent of whom are in Asia and became blind as children through poor nutrition. A small amount of appropriate foods would have saved them from a life of darkness.

And while this happens to people, the rain forests are felled, the deserts expand and dust bowls are created out of fertile land. The Amazon and other rain forests remove carbon dioxide and give us oxygen in return. They hold a thousand tons of biomass on each hectare, produced from solar energy trapped by their leaves. The water which, helped by the shade of the trees, is retained in the soil is drawn up by their roots and transpires through their leaves with such effect that it is thought that 200 billion tons of water is held in the atmosphere above Amazonia. Yet we cut them down.

According to studies produced by the Brazilian Space Research Station, adverse climatic changes are already in evidence as a result of the deforestation and topsoil is being washed into the sea to bury the sensitive coastal marine ecology. The exploitation of the timber has left bare land which could be used for farming. Tragically the residual soil is poor and a vivid BBC television programme described how the small farmers settled there can barely make a living. Those who removed the timber of these huge trees also removed the minerals.

The Sahara is marching southward at a rate of some 20 kilometres a year, an advance which produces another 2000 square kilometres of desert. Every year 150 million tons of topsoil with its minerals and fertility is blown off the misused surface of Africa south of the Sahara to land in the ocean as far away as Barbados. In August 1987 Londoners woke to find their cars covered in a fine brown dust. It was not city dirt, but it was something new. To their astonishment, they learnt it was dust – from the Sahara. At the slowest reckoning, by the end of the century the Sahara will have claimed another 250,000 square kilometres of once fertile land.

An Ethiopian peasant, remarking that a decade ago his harvest was good but since then the topsoil has been washed away, spoke for millions when he said: 'Now, all I have is a harvest of dust.' (Alan McGregor, *The Times*, 2 July 1984.) The tragedy was no doubt increased by the slow pace at which land which at first had seemed promising for grazing or cash-cropping gradually deteriorated. Sudan and Bangladesh also deserve a mention as manmade disaster areas.

Dr John Boardman, deputy director of the Countryside Research Unit in Brighton, commented on the same process in the UK (*Guardian*, 18 December 1987). Soil loss from arable land is reaching unprecedented levels with one nine-hectare plot losing this autumn (1987) a record of 270 tonnes per hectare. 'The present farmers are farming the remnants of a once rich soil which was lost through ignorance.' And he quotes Professor Morgan of Silsoe College: 'It is doubtful if sustained use of erodable areas for arable production can be maintained far into the next century . . .'

While there are long range climatic forces in operation, man visibly played his part in deforestation and desert formation. In the rain forests of the Amazon and Thailand, the deforestation, many experts believe, has gone too far. In Africa, south of the Sahara, bush clearance for grass and bore holes to reach fossil water, are being introduced to make pastures for non-indigenous cattle. An African ecology, forged over 50 million years to meet the requirements of hot, dry conditions, is being replaced by a system adapted to moist, temperate Europe. The principle is the same. And it is no different to the principle described by Plato: 'The rich, soft soil has all run away leaving the land nothing but skin and bone.'

The inexorable march in the wrong direction is not for want of contemporary warning. Sir Frank Fraser-Darling wrote at length about the issue in the middle of this century. The Club of Rome and the Brandt Report both insisted that these issues are global and demand that man accepts his responsibility for the planet. When the Minoans chopped down their first oak tree, when the groundnut, sweet potato and plantain were introduced to Africa, no one could foresee the outcome. Now that we know, we can respond. And similarly, we can respond to our own mistakes.

One simple error which needs immediate rectification is that principles of agriculture have been derived from principles of quantity, and nutrition has had no part except by accident; take, for example, the way in which the protein and nutrient value of animal products has been diluted by fat. In the European Common Agricultural Policy any connection between agriculture and health is almost as invisible as its economic reality. The latest buzzword is the harmonisation of policy. With Greece, Portugal and Spain now joining the EEC, the danger is that their insignificant death rate from heart disease will be 'harmonised' upwards in line with the UK and Northern Europe!

In Darwin's day, the evidence that man was encroaching on the most fertile lands had already been embodied a century beforehand in economic theory. Adam Smith recognised that it was the most fertile lands which were first occupied. The encouragement given by highly fertile soil led to population growth but the housing needed was built on the same land. So progressively more peripheral land had to be used that was harder to win and probably less fertile.* This process was embodied in Adam Smith's Law of Diminishing Returns.

If the economic implications of progressive change in soil fertility

* An excellent example is the city of London and its suburbs which now smothers the most fertile plain in the Thames valley. Ironically, London is now surrounded by a green belt created by law and on which no one may build.

were clear to Adam Smith they were not a topic of conversation in Darwin's time and he voiced the general feelings of the time when he wrote: 'Man has no power of altering the absolute conditions of life; he cannot change the climate of any country: he adds no new element to the soil . . . It is an error to speak of man tampering with nature . . .' In the same way, Darwin did not recognise that he was observing a last struggle for survival of species, many of which have become extinct since he wrote. Today man's 'tampering with nature' is common conversation.* It is never too late to learn, but the arrogance that grew out of the Victorian self-perception of the 'fittest' has impeded the understanding of man's relationship with the planet in the same way that religion held scientific and biological progress at a standstill. The economists lost contact with the simple message of Adam Smith. For our future success, economic theory must accept its biological and planetary responsibility. Gebbie would say that as far as their 'planet' of Australia was concerned, the Aborigines understood: we don't.

There are no more continents to discover, existing resources are limited and being used at phenomenal rates. The problem is that *Homo sapiens* has yet to reach adulthood. In his relationship with his parent planet, *Homo sapiens* is still a child that has not learnt to look after his pocket money. The rich countries are no better or worse than the poor ones. They know about vitamin A blindness: we know about saturated fats and heart disease. We can forecast the logical conclusions.

We in the Western world can only just afford elaborate and expensive health services, and by this time we should surely have realised that nutrition must have an important place in anything that is truly a health service and not merely a machine for patching up patients who have become diseased. This means first of all that our doctors need to understand the issues and that the latest findings of nutritional science must play an important part in their education.† The real facts are flabbergasting. In 1980, over the whole

* See *Time* 19 October 1987 'The heat is on' – a discussion on man tampering with the climate (the greenhouse effect) and the ozone layer.

† Some of the subject matters in the medical curriculum in which nutrition plays (or rather should play!) a prominent role are: prevention and treatment of heart disease, diabetes, cystic fibrosis, hypertension, recovery from trauma, management of liver disease and renal failure, pregnancy and lactation, obesity, anorexia nervosa, multiple sclerosis and stroke; cancer therapy (especially in recovery from chemo or radiotherapy) and the interaction between the immune system and nutrition are subjects of increasing interest.

course of their training, British medical students were given just three hours of instruction in nutrition. In the USA the situation is not much better. Yet every medical school has its full department of pharmacology to teach about the use of drugs. Fortunately, the situation is changing.

The problem is that common sense tells the public to recognise the significance of nutrition but they have few leaders. They must interpret for themselves the little information that comes from the few academic centres involved in research. The result is that everyone then has to become his own expert. Without proper science the end product is confusion: yet at the same time it is precisely that 'people response' which has achieved the most.

What is important for individuals is still more important for societies. A French philosopher wrote that the riches of a nation lay not in its iron or mineral deposits but in its intellectual wealth. In the way that the underprivileged are significantly more likely to experience foetal growth retardation and to give birth to handicapped children, so nutritional forces can manipulate whole populations, producing systematic change in a particular direction. Had the cretins of the Alps not been discovered and cured, the fossil record, stumbled across in some future epoch, would surely have described them as a different species!

We know that the average height in the UK increased by 0.4 inch per decade during the first part of the century having, according to data from the Hythe burial grounds, been fairly stable since the 13th century. IQ assessments of children have shown that those born to poor families fare worse than those who are well off: the cycle of deprivation. The cross-cultural data also shows that Japanese children are scoring at a significantly higher level than their American counterparts. The Westinghouse Science Merit Awards in the USA have been won over the last five years mainly by children of Japanese or Far Eastern origin. At the other end of the scale, children are born never to achieve at school in far larger numbers from the lower socio-economic groups. Some 10 per cent of children in the UK require additional educational attention. We know that there are strong nutritional backgrounds to these situations. What if we could eliminate the bottom end of deprivation: would not the average intelligence increase?

If humans are growing taller (and perhaps more athletic) and if this is associated with different nutrition extending over but a few generations, could there be a similar improvement in human intelligence? Theoretically, there is no reason why not: if nutrition can change the shape and size of the individual's body at a speed far faster than selective processes could hope to operate, is there any reason why the brain may not also be affected? We have ample examples which tell us that brain size can diminish with an increase in body size. This question is too important to dodge.

We would not want to deliberately decrease brain capacity but this could happen if we are not careful. The extraordinary fact is that we could actually define the conditions which should at least eliminate low birth weight, foetal growth retardation and handicap of non-genetic origin. This in itself would have to improve average intelligence. But could we lift the mean? Could we achieve higher intelligence in the way that we run faster and jump higher?

In practice, an approach based on conserving or improving physical and intellectual health demands a multidimensional approach. Nutrition is not just a set of individual drug-like vitamins: it is multifactorial. Food is several things but different foods provide different groups or clusters of nutrients. It is not just the single nutrient that matters: the relationship between them matters too. Different animal species have different requirements and in the mammals this truth is reflected in the composition of their milks which are rich in protein on the one hand for fast body growth or rich in essential fatty acids on the other, when the postnatal focus is on brain growth. In the human species the highest specialisation, which stands out head and shoulders above other species, is the brain and it is built in the womb of the mother. We must also remember that the other, interwoven specialisation, the vascular system, grows when the child is growing. We also need to remember the universal rule that any system is most vulnerable to external influences during its development and growth.

The challenge we face is clear: it is to stop the degenerative diseases and, instead, to continue the evolution and development of the human brain and intelligence.

In addition to considerations of nutrition and health, success requires giving education a far higher priority than it presently receives. To respond to this challenge requires knowledge, particularly on the part of women, whose bodies and minds hold the responsibility for the next generation. Yet in our present day world, two-thirds of women are illiterate and only 1 per cent own property.

One highly placed WHO worker commented that if he had a choice, he would give education to the girls in preference to the boys. In a future world, the education of women must assume a high priority.

On the biological side we must eliminate malnutrition and ensure that all female children, not just a favoured few, understand the importance of and have available a plentiful supply of education and good foods, especially those important to the nervous and vascular tissue. This will only have its full effect when children are born to mothers who have, from conception, been similarly nourished. To bring that about we need to build a world society willing to share its resources and control its populations to eliminate the miseries of poverty, hunger and malnutrition, and somehow we have to find more creative outlets for human energy than the wars and other forms of aggression that throughout history have called on man's greatest resources of strength and ingenuity. We need to devote that energy instead, to works of imagination, art, science and education, and we need above all to understand that the care of our environment is not an optional extra, a harmless pastime for the well fed middle classes, but the central task for our age. If we fail in that we fail in everything, for we are a part of that environment. Artificial intelligence may come to our aid and give us a breathing space, fending off catastrophe long enough. But we must not give undue importance to artificial intelligence, for it is simply an extension of human intelligence and an aid for it.

It should now be obvious that biological considerations ought to direct agricultural policies world-wide, yet in reality the aim of 'agricultural development' is often pursued for different reasons. At the same time, the story is far from being all doom and misery. The advances in science and technology in this century alone leave one with no alternative view other than that man has the ability to do the right thing.

If society does respond to the WHO call of 'Health for All', and if we do accept the challenge that the shape of human development is in our own hands, then the mind can only race through the most exciting images of the world in 50 years from now. The achievements since the turn of the century when Marconi flew a box kite above Signal Hill in Newfoundland and heard that first transatlantic radio message in the form of three faint dots, are extraordinary. In this short space of time man has gone from box kite to satellite.

The tragedy is that man's progress in ecology and land use has been in the opposite direction. It is up to our present generations to take the necessary actions and decide the direction of our own evolution and the shape of our future. The only downside of the equation is that for all the brilliant advances in science and technology, the same care to detail and excellence has not been devoted to our use of land and marine resources. Indeed, the pollution of the rivers, lakes, oceans, the deforestation, desert formation and famine all speak of the opposite.

In learning to decide wisely we would be greatly helped by a new understanding of human evolution to extend that of Darwin and supplement it with insights that the Darwinian view cannot itself provide. It is not only that the theory we have described in this book can give us predictive power in scientific and practical terms: it also gives us a wholly different philosophical approach. Are we what we are because of a long process of selection involving the unceasing war of all against all in which the survivors demonstrate their 'fitness' by wiping out their rivals? Or are we instead the product of a more generous world, whose children are called into existence by the wealth of new opportunities which at certain critical moments she presents to them? Were we shaped by a concatenation of random events, unpredictable and unpatterned, or are we rather the latest creation of a universe governed everywhere and always by the laws of its own development; laws which through science we, its creatures, can hope to understand and to use in shaping our own destiny?

That is the image of the world that we believe to be the true one. If we look at evolution we can see a long history but not a disorderly one. When the elements were first forged in the great star-furnaces and supernovae, what came out of that process were not random combinations of fundamental particles but just those few particular ones that the nature of matter dictates. When compounds of elements formed on the cooling earth they too were predetermined by the laws of matter, and so in due course were the organic chemicals and the biochemicals that, when the conditions were right, shaped themselves into the first biotic matter and then into the first true living cells again in the shape and form that chemistry and nature allows. From those primal cells to the plants, the animals, the mammals and at last to man himself, the whole process has been one of ordered interactions between organisms and their environment.

Man is one of many creatures that long progression has thrown up, and not in all ways the most favoured or the best adapted. What can give him hope is that he alone can look back and see what made him, and look forward to gauge what that knowledge implies for his future. Among that knowledge is a fact so simple that it is strange it should be so often ignored: whatever else he may be, man is a particular structure of organic chemicals. To maintain that structure in good order the materials of which it is built – that is to say its foods – must be right. Above all they must be right during that critical time when the organism is being formed in the womb, to emerge in due course as one of a new generation of that species whose actions will determine the fate of every species on earth. Standing at the moment of intersection between the vast gulf of the past and the dangers and promises of the future there stands a figure whose importance overwhelms all others: the figure of a human mother and child.

Bibliography

Ackman, R. G. 1982. Fatty acid composition of fish oils. In *Nutritional Evaluation of Long-chain Fatty Acids in Fish Oil*, ed. S. M. Barlow & M. E. Stansby, London, Academic Press, 25–88

Ackman, R. G. 1989. In *The Role of Dietary Fats in Human Nutrition*, ed. A. Vergroessen & M. A. Crawford, London, Academic Press

Adams, J. M. & Cory, S. 1983. Immunoglobin genes. In *Eukaryotic Genes*, ed. N. Maclean, S. P. Gregory & R. A. Flavell, London, Butterworth, 343–58

Allsopp, A. 1969. Phylogenetic relationships of the prokaryotes and the origin of the eukaryotic cell. In *New Phytologist*, **68**, 591–612

Althabe, O. & Laberre, C. 1985. Chronic villitis of unknown aetiology and intrauterine growth-retarded infants of normal and low ponderal index. In *Placenta*, **6**, 265–276

Althabe, O., Laberre, C. & Telenta, M. 1985. Maternal vascular lesions in placentae of small for gestational age infants. In *Placenta*, **6**, 265–76

Amoore, J. E. 1961. Dependence of mitosis and respiration in roots upon oxygen tension. *Proc. R. Soc. Lond. B*, **154**:109–116

Anders, E. & Owen, T. 1977. Mars and Earth: origin and abundance of volatiles. *Science*, **198**:453–465

Anderson, H. T. 1966. Physiological adaptations in diving vertebrates *Physiol. Rev.*, **46**:109

Anderson, R. E., Benolken, R. M., Jackson, M. B. & Maude, M. B. 1977. The relationship between membrane fatty acids and the development of the rat retina. In *Functions and Biosynthesis of Lipids*, ed. N. G. Bazan, R. R. Brenner, N. M. Guisto, 547–559, New York and London, Plenum Press

Angel, H. 1985. *Life in the Oceans*, London, Treasure Press

Ardrey, R. 1961. *African Genesis*. London, Collins

Aristotle, *Meteorologica*, Book 1, Ch.14. Meteorological Institute of the University of Thessalonica, 1970–7

Asmussen, I. 1978. Arterial changes in infants of smoking mothers. *Post Grad. Med. J.*, **54**:200–204

Awramik, S. M. & Barghoorn, E. S. 1977. The gunflint microbiota. *Precambrian Res.*, **5**:121–143

Bakker, R. T. 1978. Dinosaur feeding behaviour and the origin of flowering plants, *Nature*, **274**:661

Baldwin, E. 1949. *An Introduction to Comparative Biochemistry* (3rd edn) Cambridge, Cambridge University Press
Baldwin, J. M. 1902. *Development and Evolution: Including Psychophysical Evolution, Evolution by Orthoplasy and the Theory of Genetic Modes*. New York, Macmillan
Bang, H. O., Dyerberg, J. & Sinclair, H. M., 1980. The composition of Eskimo food in North Western Greenland. In *Amer. J. Clin. Nutr.*, **33**:2657
Banks, H. P. 1970. Major evolutionary events and the geological record of planets. In *Biol. Rev.*, **47**:451–454. Cambridge
Barghoorn, E. S. 1971. The oldest fossils. In *Scientific American*, **224**:30–42
Barghoorn, E. S. 1974. Two billion years of prokaryotes and the emergence of eukaryotes. In *Taxon*, **23**:259
Barker, D. J. P. & Osmond, C. 1986. Infant mortality, childhood nutrition and ischaemic heart disease in England and Wales. In *Lancet* **i**:1077–81
Barker, D. J. P. and Osmond, C. 1987. Inequalities in health in Britain: specific explanations in three Lancashire towns. In *Br. Med. Journal*, **294**:749–52
Bassham, J. A. & Jensen, R. G. 1967. Photosynthesis of carbon compounds. In *Harvesting the Sun* ed. A. San Pietrol, T. Greer & J. Army, New York, Academic Press, 79–110
Bateson, W. 1894. *Materials for the Study of Variation*. London, Macmillan
Bazan, N. G. 1989. Lipid derived metabolites as possible messengers: arachidonic acid, leukotrienes, docosanoids, and platelet activating factor. In *Extracellular and Intracellular Messengers in the Vertebrate Retina*, Alan R. Liss Inc., 269–300
Beisson, J. & Sonneborn, T. M. 1965. Cytoplasmic inheritance of the organisation of the cell cortex. *Paramecium aurelia, Proc. Natl. Acad. Sci. USA.*, **53**:265–82
Bellamy, D. 1978. *Botanic Man*, London, Hamlyn
Bender, A. 1980. *The Role of Plants in Feeding Mankind* Frey Ellis Memorial Lecture, Vegan Society, 33–35 George Street, Oxford OX1 2AY
Bender, W. 1985. Homeotic gene products as growth factors. *Cell*, **43**:559–60
Benemann, J. R. 1973. Nitrogen fixation in termites. *Science*, **181**:164–5
Bernal, J. D. 1967. *Origin of Life*, Cleveland, Ohio, World
Bernfield, M. R. 1981. Organisation and remodelling of the extracellular matrix in morphogenesis. In *Morphogenesis and Pattern Formation*, ed. T. G. Connelly, L. L. Brinkley & B. M. Carlson, New York, Raven Press, 139–62
Bernard, C. 1978. *Claude Bernard on the internal environment: a memorial symposium*, ed. Eugene Debs Robin, New York, Basel, Decker, c1979. Proc. Symp. Stanford Univ

Birch, H. & Gussow, J. 1970. *Disadvantaged Children, Health, Nutrition and School Failure*. New York, Harcourt, Brace & World

Bishop, J. M. 1983. Cellular oncogenes and retroviruses. In *Ann. Rev. Bichem.*, **52**:301–54

Bjerve, J. D., Mostad, I. L. & Thoresen, L. 1987. Alpha-linolenic acid deficiency in patients on long-term gastric tube feeding: estimation of linolenic acid and long chain unsaturated n–3 fatty acid requirement in man. In *American Journal of Clincial Nutrition*, **45**:66–77

Blau, H. M., Parlath, G. K., Hardeman, E. C., Chin, C. P., Silberstein, L., Webster, S. G., Miller, S. C. & Webster, C. 1985. Plasticity of the differentiated state. In *Science*, **230**:758–66

Blaxter, K. & Waterlow, J. (eds) 1985. *Nutritional Adaptation in Man*. London, J. Libbey

Bohinski, R. C. 1987. *Modern Concepts in Biochemistry* (5th edn) Boston, Allyn Bacon Inc

Borgens, R. B. 1982. What is the role of naturally produced electric current in vertebrate regeneration and healing? In *Int. Rev. Cytol.*, **76**:245–98

Bowen, I. D. & Lockshin, R. A. 1981. *Cell Death in Biology and Pathology*. London, Chapman and Hall

Bowler, P. 1983. *The Eclipse of Darwinism*. Baltimore, John Hopkins University Press

Bowler, P. J. 1975. The changing meaning of evolution. *J. Hist. Ideas* **36**:95–114

Bowler, P. J. 1979. Theodor Eimer and orthogenesis: evolution by definitely directed variation. In *J. Hist. Medicine* **34**:40–73

Brennan, C. & Winet, H. 1977. Fluid mechanics of propulsion by cilia and flagella. In *Ann. Rev. Fluid Mech.*, **9**:339–98

Brick, I. 1966. Circulatory response to immersing the face in water. In *J. Applied Physiol.*, **21**:540

Broda, E. 1975. *The Evolution of the Bioenergetic Process*. Oxford, Pergamon Press

Brown, D. D. 1981. Gene expression in eukaryotes. In *Science*, **211**:667–74

Brown, R. E., Shaffer, R. D., Hansen, I. L., Hansen, H. B. & Crawford, M. A. 1966. Health survey of the El Molo. In *E. Afr. Med. J.*, **43**:480–8 *et seq*

Brun, R. B. 1978. Developmental capacities of Xenopus eggs provided with erythrocyte or erythroblast nuclei from adults. In *Devel. Biol.*, **65**:277–84

Bruno, J., Reich, N. & Lucas, J. J. 1981. Globin synthesis in hybrid cells constructed by transplantation of dormant avian erythrocyte nuclei into enucleated fibroblasts. In *Mol. Cell. Biol.*, **1**:1163–76

Bryant, P. J. & Simpson, P. 1984. Intrinsic and extrinsic control of growth in developing organs. In *Quart. Rev. Biol.*, **59**:387–415

Buchsbaum, R. 1968. *Animals without Backbones: 2* London, Pelican Books

Budowski, P. & Crawford, M. A., 1985. Alpha-linolenic acid as a regulator of the metabolism of arachidonic acids: dietary implications of the ratio n–6:n–3 fatty acids. In *Proc. Nut. Soc.*, **44**:221–9

Budowski, P., Leighfield, M. J. & Crawford, M. A. 1987. Nutritional encephalomalacia in the chick: an exposure of the vulnerable period for cerebellar development and the possible need for both w6 and w3 fatty acids. In *Br. J. Nutr.*, **58**:511–20

Burnet, F. M. 1959. *The Clonal Selection Theory of Acquired Immunity*. Nashville, Vanderbilt University Press

Burr, G. O. & Burr, M. M. 1930. On the nature and role of the fatty acids essential in nutrition. In *J. Biol. Chem.*, **86**:587–621

Cairns, J. *et al*. 1988. The origin of mutants. In *Nature*, **3235**:142–5

Calder, W. A. 1978. The Kiwi. In *Sc. Amer.*, **239(1)**:102

Calvin, M. 1969. *Chemical Evolution: Molecular Evolution towards the Origin of Living Systems on the Earth and Elsewhere*. New York, Oxford University Press

Carell, E. F. 1969. Studies on chloroplast development and replication in Euglena. 1. Vitamin B12 and chloroplast replication. In *J. Cell Biol.*, **41**:431–40

Carroll, K. K., Hopkins, G. J., Kennedy, T. G. & Davidson, M. B., 1981. Essential fatty acids and cancer. In *Progr. Lipid Res.*, **20**:685–90

Cavalier-Smith, T. 1975. The origin of nuclei and of eukaryotic cells. In *Nature*, **256**:463–7

Cavalier-Smith, T. 1978. The evolutionary origin and phylogeny of microtubules, mitotic spindles and eukaryotic flagella. In *Biosystems*, **10**:93–114

Chada, K., Magram, J. & Constantini, F. 1986. An embryonic pattern of expression of a human foetal globin gene in transgenic mice. In *Nature*, **319**:685–9

Clandinin, M. T., Chapell, J. E. & Heim, T. 1981. Do low weight infants require nutrition with chain elongation-desaturation products of essential fatty acids? In *Prog. Lipid Res.*, **20**:901–04

Clark, G. & Piggott, S. 1965. *Prehistoric Societies*, London, Hutchinson

Clarkson, S. G. 1983. Transfer RNA genes. In *Eukaryotic Genes: Structure, Activity and Regulation*, ed. N. Maclean, S. P. Gregory & R. Flavell, London, Butterworth, 239–62

Cleave, T. L. 1974. *The Saccharine Disease*, Bristol, John Wright

Cloud, P. 1976. Beginnings of biospheric evolution and their biogeochemical consequences. In *Paleobiology*, **2**:351–87

Cloud, P. E. Jr. 1968. Premetazoa evolution and the origin of metazoa. In *Evolution and Environment*, ed. E. T. Drake, New Haven, Yale University Press, 1–72

Cloud, P. E. Jr. 1974. Evolution of ecosystems. In *Am. Sci.*, **62**:54–66

Cloudsley-Thompson, J. 1974. *Desert Life*. London, Aldus Books

Clutton-Brock, J. 1981. *Domesticated Animals from Early Times*. London, British Museum of Natural History

Colinvaux, P. 1974. *The Fates of Nations*. London, Pelican
Committee on Diet, Nutrition & Cancer. 1982. *Assembly of Life Sciences, National Research Council*. Washington, DC, National Academy Press
Cooper, M. & Pinkus, H. 1977. Intrauterine transplantation of rat basal cell carcinoma: a model for reconversion of malignant to benign growth. *Cancer Res.*, **37**:2544–52
Coronary Prevention Group. 1978. The prevention of heart disease starts in childhood. In *Post Grad. Med. J.*, **54**:137–230
Crawford, M. A. 1968. Fatty acid ratios in free-living and domestic animals. In *Lancet* **i**, 1329–33
Crawford, M. A. 1976. *The Living Earth: Conservation*. New York, Danbury Press
Crawford, M. A. 1986. Heart disease and cancer: are different diets necessary? In *Diet and the Prevention of Coronary Heart Disease and Cancer*. Ed. B. Hallgren, E. Rossner, IVth International Berzelius Symposium, Stockholm. New York, Raven Press, 149–58
Crawford, M. A., Casperd N. M., & Sinclair A. J. 1976. The long-chain metabolites of linoleic and linolenic acids in liver and brain in herbivores and carnivores. In *Comp. Biochem. Physiol.*, **54B**:395–401
Crawford, M. A. & Crawford, S. M. 1972. *What We Eat Today*. London, Neville Spearman
Crawford, M. A., Crawford, S. M. & Berg Hansen, I., 1970. Plasma structural lipids in groups at high and low risk of atherosclerosis. In *Biochem. J.*, **122**:11–12
Crawford, M. A., Doyle, W., Drury, P., Leighfield, M., Lennon, A., Costeloe, K. & Leaf, A. 1989. n–6 and n–3 fatty acids during early human development. In *J. Int. Med.*, **225**:159–69
Crawford, M. A., Doyle, W., Craft, I. L. & Laurance, B. M. 1986. A comparison of food intakes during pregnancy and birthweight in high and low socio-economic groups. In *Prog. Lipid Res.*, **25**:249–54
Crawford, M. A., Gale, M. M. & Woodford, M. H. 1969. Linoleic acid and linolenic acid elongation products in muscle tissue of Syncerus caffer and other ruminant species. In *Biochem. J.*, **115**:25–7
Crawford, M. A., Gale, M. M. & Woodford, M. H. 1970. Muscle and adipose tissue lipids of the Warthog (*Phacochoerus aethiopicus*). In *Int. J. Biochem.*, **1**:654–58
Crawford, M. A., Gale, M. M., Woodford, M. H. & Casperd, N. M. 1970. Comparative studies on fatty acid composition of wild and domestic meats. In *Int. J. Biochem.*, **1**:295–305
Crawford, M. A., Hall, B., Laurance, B. M. & Munhambo, A. 1976. Milk lipids and their variability. In *Curr. Med. Res. Opinion* (Suppl.1) **4**:33–43
Crawford, M. A., Hansen, K. L. & Lopez, A. 1969. The excretion of 3-hydroxyanthranilic and quinolinic acid in Uganda Africans. In *Br. J. Cancer*, **23**:644–54
Crawford, M. A. & Sinclair, A. J. 1972. Nutritional influences in the

evolution of the mammalian brain. In *Lipids, Malnutrition and the Developing Brain*, ed. K. Elliot, J. Knight. A Ciba Foundation Symposium, Amsterdam, Elsevier, 267–92

Crawford, M. A., Hassam, A. G., Stevens, P. A. 1981. Essential fatty acid requirements in pregnancy and lactation with special reference to brain development. In *Prog. Lipid Res.*, **20**:30–40

Crawford, S. M. & Crawford, M. A. 1974. An examination of systems of management of wild and domestic animals based on African ecosystems. In *Animal Agriculture*, ed. H. H. Cole & M. Ronning, San Francisco, W. H. Freeman, 218–34

Cronin, J. E., Boaz, N. T., Stringer, C. B. & Rak. Y. 1981. Tempo and mode in hominid evolution. In *Nature*, **292**:113–22

Cunnane, S. C. 1980. The aquatic ape theory reconsidered. In *Medical Hypotheses*, **6**:49–58

Cunnane, S. C., Keeling, P. W. N., Thompson, R. P. H. & Crawford, M. A. 1984. Linoleic acid and arachidonic acid metabolism in human peripheral blood leucocytes: comparison with the rat. In *Brit. J. Nut.*, **51**:209–17

Crick, F. H. C. 1968. The origin of the genetic code. In *J. Mol. Biol.*, **38**:367–79

Cronquist, A. 1968. *Evolution and Classification of Flowering Plants.* Boston, Houghton Mifflin

Dam, H., Neilsen, G. K., Prange, I. & Sondergaard, E. 1958. The influence of linoleic and linolenic acids on symptoms of vitamin E deficiency in chicks. In *Nature*, **182**:802–03

Darby, W. J., Ghalioungui, P. & Grivetti, L. 1977. *Food: The Gift of Osiris*, London, Academic Press

Darlington, C. D. 1969. *The Evolution of Man and Society*, London, Allen & Unwin

Darwin, C. 1842. *The Foundations of the Origin of Species, a sketch written in 1842*. ed. Francis Darwin, Cambridge, CUP, 1909

Darwin, C. 1845. *A Naturalist's Voyage Round the World*, London, John Murray

Darwin, C. 1959. *The Voyage of the 'Beagle'*. Everyman's Library, vol.104. London, J. M. Dent & Sons Ltd

Darwin, C. 1868. *The Origin of Species by Means of Natural Selection: or the Preservation of Favoured Races in the Struggle for Life*, London, John Murray

Darwin, C. 1871. *The Descent of Man, and Selection in Relation to Sex*, London, John Murray

Darwin, C. 1876. *Autobiography*, London, John Murray

Davidson, E. H. 1968. *Gene Activity in Early Development*, New York, Academic Press

Davidson, E. H. & Britten, R. J. 1973. Organization, transcription and regulation in the animal genome. In *Quart. Rev. Bio.*, **48**:565–613

Davie, R., Butler, N. R., Goldstein, H. 1972. *From birth to seven: the*

2nd report of the National Child Development Survey, London, Longman

Davies, J. N. P., Knowelden, J. & Wilson, B. A. 1965. Incidence rates of cancer in Kayadondo County, Uganda, 1954–1960. In *J. Nat. Cancer Inst.*, **35**:789–821

Dawber, T. R. 1980. *The Framingham Study: The epidemiology of atherosclerotic disease*. Cambridge MA, Harvard University Press

Dawkins, R. 1976. *The Selfish Gene*, Oxford, Oxford University Press

Dawkins, R. 1986. Creation and natural selection. In *New Scientist*, **25**:34–8

Dawkins, R. 1986. *The Blind Watchmaker*, London, Longman

Degens, E. T. 1976. Molecular mechanisms of carbonate, phosphate and silica deposition in the living cell. In *Top. Curr. Chem.*, **64**:3–112

Denton, G. H. and Hughes, T. J. (eds) 1981. *The Last Great Ice Sheets*, New York, John Wiley

De Jong, W. W. *et al*. 1981. Relationship of aardvark to elephants, hyraxes and sea cows from alpha-crystalline sequences. In *Nature*, **292**:538

De Vries, H. 1906. *Species and Varieties: Their Origin by Mutation*, Chicago, Open Court

De Vries, H. 1910. *Intracellular Pangenesis: Including a Paper on Fertilisation and Hybridisation*, Chicago, Open Court

Department of Health and Social Security. 1974. Diet and Coronary Heart Disease: report of the Advisory Panel of the Committee on Medical Aspects of Food Policy on diet in relation to cardiovascular disease and cerebrovascular disease. In Report on Health and Social Subjects, London. HMSO, 7

Department of Health and Social Security. 1984. Report on Health and Social Subjects 28, Diet and Cardiovascular Disease. *Committee on Medical Aspects of Food Policy*, London, HMSO

Dickerson, R. E. 1980. Cytochrome C and serendipity. In *Evolution of Protein Structure and Function*, ed. D. S. Sigman, M. A. B. Brazier, UCLA Forum in Medical Science 21, New York, Academic Press

Dimroth, E. & Kimberley, M. 1976. Precambrian atmospheric oxygen: evidence in the sedimentary distributions of carbon, sulfur, uranium, and iron. In *Can. J. Earth Sci.*, **13**:1161–85

Dobbing, J. 1972. Vulnerable periods of brain development. In *Lipids, Malnutrition & the Developing Brain*, ed. K. Elliott, & E. Knight, Amsterdam: Associated Scientific Publishers, A Ciba Foundation Symposium, Amsterdam, Elsevier, 9–22

Dobson, E. O. 1979. Crossing the prokaryote-eukaryote border: endosymbiosis or continuous development? In *Can. J. Microbiol.*, **25**:652–74

Dodge, O. G. & Linsell, C. A. 1963. Carcinoma of the penis in Uganda and Kenya Africans. In *Cancer*, **16**:1255–63

Doll, R., Muir, C. & Waterhouse, J. 1968. *Cancer Incidence in Five Countries*. New York, Springer-Verlag

Dormandy, T. L. 1988. In praise of peroxidation. In *Lancet* **ii**:1126–8
Doyle, W., Crawford, M. A. & Laurance, B. M. 1982. Dietary survey during pregnancy in a low socio-economic group. In *Journal of Human Nutrition*, **36a**:95–106
Drummond, J. C. & Wilbrahim, A. 1969. *The Englishman's Food*. London, Jonathan Cape
Du Boulay, G. H. & Crawford, M. A. 1968. Nutritional bone disease in captive primates. *Comparative Nutrition of Wild Animals*, ed. M. A. Crawford, *Symp. Zool. Soc. London*, **21**:223–36
Dyerberg, J., Bang, H. O., Stofferson, E., Moncada, S., Vane, J. R. 1978. Eicosapentaenoic acid and the prevention of thrombosis and atherosclerosis. In *Lancet*, **i**:117–9

Eaton, S. B. & Konner, M. 1985. Paleolithic nutrition. In *New England Journal of Medicine*, **312, no5**:283–9
Echlin, P. 1970. The photosynthesis apparatus in prokaryotes and eukaryotes. In *Organization and Control in Prokaryotes and Eukaryotes*. Ed. H. P. Charles & B. D. Knight, *Symp. Soc. Gen. Microbiol.*, Cambridge, Cambridge University Press, **20**:221–48
Ehrlich, P. & Ehrlich, A. H. 1970. *Population Resources: Environment*. San Francisco, W. H. Freeman
Eldredge, N. 1986. *Time Frames: Rethinking of Darwinian Evolution and the Theory of Punctuated Equilibria*, London, Heinemann
Eldredge, N. 1986. *Unfinished Synthesis: Biological Hierarchies and Modern Evolutionary Thought*, New York, Oxford University Press
Essential Fatty Acids and Prostaglandins: Progress in Lipid Research, ed. R. T. Holman, **21** (1981) and **25** (1986)
Etkin, L. D. & DiBerardino, M. A. 1983. Expression of nuclei and purified genes microinjected into oocytes and eggs. In *Eukaryotic Genes*, ed. N. Maclean, S. P. Gregory & R. Flavell, London, Butterworth, 127–56

FAO/WHO 1978. The Role of Dietary Fats and Oils in Human Nutrition. Conjoint Expert Consultation, *Nutrition Report No 3* FAO, Rome
Fiennes, R. T. N. W. (ed.) 1972. Biology of Nutrition. In *International Encyclopaedia of Food and Nutrition*, Oxford, Pergamon Press
Finn, P. J., Nielsen, N. O. 1971. The inflammatory response of rainbow trout. In *J. Fish Biol.*, **3**:463
Fisher, S., Weber, P. C., 1983. Prostaglandin I3 is formed *in vivo* in man after dietary eicosapentaenoic acid. In *Nature*, **307**:165–8
Foe, V. E., Wilkinson, L. E. & Laird, C. D. 1976. Comparative organization of active transcription units in *Oncopeltus fasciatus*. In *Cell*, **9**:131–46
Fox, S. W. 1965. *The Origin of Pre-biological Species*, New York, Academic Press

Fox, S. W. & Harda, K. 1960. Thermal copolymerisation of amino-acids common to proteins. In *J. Amer. Chem. Soc.*, **82**:3745–51
Fulton, C. & Dingle, A. D. 1967. Appearance of flagellate phenotype in populations of *Naefleria* amoebae. In *Devel. Biol.*, **15**:165-91
Fulton, C. & Walsh, C. 1980. Cell differentiation and flagellar elongation in *Naegleria gruberi*: dependence on transcription and translation. In *J. Cell Biol.*, **85**:346–60

Galli, C. & Socini, A. 1983. Dietary lipids in pre- and post-natal development. In *Dietary Fats and Health*, ed. E. G. Pekins & W. J. Visek, American Oil Chemists' Society, 278–301
Galli, C., Galli, G., Spagnuolo, C., Bosisio, E., Tosi, L., Folco, G. C., Longiave, D. 1977. *Function and Biosynthesis of Lipids*, ed. N. G. Bazan, R. R. Brenner & N. M. Giusto, New York, Plenum Press Ltd, 561–73
Gebbie, D. A. M. 1981. *Reproductive Biology: Descent Through Woman.* New York, John Wiley & Sons Ltd
Gey, K. F. On the antioxidant hypothesis with regard to arteriosclerosis. In *Biblthca Nutr. Dieta*, **37**:53–91
Gibbons, I. R. 1965. Chemical dissection of cilia. In *Arch. Biol. (Liege)*, **76**:317–40
Gibbons, I. R. & Rowe, A. J. 1965. Dynein: a protein with adenosine triphosphatase activity from cilia. In *Science*, **149**:424
Gilbert, S. F. 1985. *Developmental Biology*. Sunderland, Mass. Sinauer Associates Inc
Giusto, J. P. & Margulis, L. 1981. Karyotypic fissioning theory and evolution of old world monkeys and apes. In *Biosystems*, **13**:200–50
Glaessner, M. F. 1984. *The Dawn of Animal Life: A Biohistorical Study*. Cambridge Earth Science Series, Cambridge University Press
Glaessner, M. F. 1968. Biological events and the precambrian time scale. In *Can. J. Earth Sci.*, **5**:585–90
Goldschmidt, R. 1933. Some aspects of evolution. In *Science* **78**:539–47
Goldschmidt, R. 1940. The material basis for evolution, (reprint, Paterson, N. J., Pageant Books, 1960) In *The Eclipse of Darwinism*, ed. P. J. Bowler, Paterson, N. J., Pageant Books
Goldschmidt, R. 1949. Research and politics. In *Science*, 4 March 1949
Goss, R. J. 1964. *Adaptive Growth*. New York, Academic Press
Gould, S. J. 1975. Allometry in primates, with emphasis on scaling and the evolution of the brain. In *Contrib. Primatol.*, **5**:244–92
Gould, S. & Eldredge, N. 1977. Punctuated equilibrium; the tempo and mode of evolution considered. In *Paleobiology*, **3**:115–51
Gould, S. J. 1982. Punctuated Equilibrium – A Different Way of Seeing. In *New Scientist* **94**:137–41
Gould, S. J. 1983. *Hens' Teeth and Horses' Toes*, New York, W. H. Norton
Gregory, S. P., Maclean, N. & Pocklington, M. J. 1981. Artificial modification of nuclear gene activity. In *Int. J. Biochem.*, **13**:1047–63

Gribbin, J. 1982. *The Origin of Man and the Universe*. Oxford, Oxford University Press
Gribbin, J. & Cherfas, J. 1982. *The Monkey Puzzle: Are Apes Descended from Man?* London, The Bodley Head
Gurdon, J. B. & Uehlinger, V. 1966. *Nature*, **210**:1240

Hall, B. K. 1987. Tissue interactions in the development and evolution of the vertebrate head. In *Development and Evolution of the Neural Crest*, ed. P. F. A. Maderson, New York, John Wiley & Sons
Haldane, J. B. S. 1930. *The Causes of Evolution*. New York, Harper
Hamilton, W. D. 1964. The genetical theory of social behaviour (I & II). In *J. Theoretical Biol.*, **7**:I:1–116; II:17–32
Hansen, A. E., Wiese, H. F., Boelsche, A. N., Haggard, M. E., Adam, D. J. D. & Davis, H. 1963. Role of linoleic acid in infant nutrition: clinical and chemical study of 428 infants fed on milk mixtures varying in kind and amount of fat. In *Pediatrics*, **31**:171–92
Harbige, L. S., Crawford, M. A., Jones, R., Preece, A. W. & Forti, A. 1986. Dietary intervention studies on the phosphoglyceride fatty acids and electrophoretic mobility of erythrocytes in multiple sclerosis. In *Prog. Lipid. Res.*, **25**:243–8
Hardy, A. 1960. Was man more aquatic in the past? In *New Scientist*, 642–5
Hardy, A. 1960. Will man be more aquatic in the future? In *New Scientist*, 730–3
Hardy, A. 1965. *The Living Stream*, London, Collins
Hardy, M. H. 1983. Vitamin A and the epithelialmesenchymal interactions in skin differentiation. In *Epithelial-Mesenchymal Interactions in Development*, ed. R. H. Sawyer & J. F. Fallon, New York, Praeger Publishers, 163–88
Harrison, J. W. H. and Garrett, F. C. 1925–6. The induction of melanism in the lepidoptera and its subsequent inheritance. In *Proc. Roy. Soc. London (B)*, **99**:241–63
Hatanaka, H. & Egami, F. 1977. Selective formation of certain amino-acids from formaldehyde and hydroxylamine in a modified sea medium enriched with molyblate. In *J. Biochem.*, **82**:499–502
Haugaard, N. 1968. Cellular mechanisms of oxygen toxicity. In *Physiol. Rev.*, **48**:311–73
Heagerty, A. M., Ollerenshaw, J. D., Robertson, D. I., Bing, R. F. & Swales, J. D. 1986. Influence of dietary linoleic acid on leucocyte sodium transport and blood pressure. In *Br. Med. J.*, **293**:295–7
Hennig, W. 1966. *Phylogenetic Systematic*, Urbana, University of Illinois Press
Heyerdahl, T. 1974. *Sea Routes to Polynesia*, London, George Allen & Unwin Limited
Hilditch, T. P. & Williams, P. N., 1964. *The Chemical Composition of Natural Fats*, London, Chapman & Hall

Hiroa, T. R. & Buck, P. 1982. *The Coming of The Maori*, Whitcoulls Limited

Hjermann, I., Byre, K. V., Holme I. & Leren, P. 1981. Effect of diet and smoking intervention on the incidence of coronary heart disease. Report from the Oslo Study Group of a randomised trial in healthy men. *Lancet*, **ii**:1303–10

Hobhouse, H. 1985. *Seeds of Change*, London, Sidgwick & Jackson

Holland, H. D. 1978. *Chemistry of the Atmosphere and Oceans*, New York, Wiley Interscience

Holman, R. T. 1977. The deficiency of essential fatty acids. In *Polyunsaturated Fatty Acids*, ed. W. Kunau & R. T. Holman, Cincinnati Proc. AOCS USA, **4**:163–82

Holman, R. T., Johnson, S. B. & Hatch, T. F. 1982. A case of human linolenic acid deficiency involving neurological abnormalities. In *Am. J. Clin. Nutr.*, **35**:617–23

Holton, R. W., Blecker, H. H. & Stevens, T. S. 1968. Fatty acids in blue-green algae: possible relationship to taxonomic position. In *Science*, **160**:545–7

Hornstra, G. & Lussenberg, R. N. 1975. Relationship between the type of dietary fatty acid and arterial thrombosis tendency in rats. In *Atherosclerosis*, **22**:499–516

Hornstra, G. 1985. Dietary lipids, platelet function and arterial thrombosis in animals and man. In *Proc. Nutr. Soc.*, **44**:371–8

Horstadius, S. & Josefsson, L. 1972. Morphogenetic substances from sea urchin eggs: isolation of animalizing substances from developing eggs of *Paracentrotus lividus*. In *Acta Embryol. Exp.*, **1**:7–23

Houslay, M. D., Stanley, K. K. 1982. In *Dynamics of Biological Membranes*, New York, John Wiley & Sons Ltd, 39–137

Hoyle, F. and Wickramasinghe, C. 1980. *The Origin of Life*, Cardiff, University College of Cardiff Press

Hsu, K. J. 1980. Terrestial catastrophe caused by cometary impact at the end of the Cretaceous. In *Nature*, **285**:402

Hunt, T. C., Rowley, A. F. 1986. Leukotriene B4 inducer enhanced migration of fish leucocytes *in vitro*. In *Immunology*, **59**:563–8

Hungate, R. E. 1966. *The Rumen and Its Microbes*, New York, Academic Press

Hurst, H. E. 1952. *The Nile*, London, Constable

Hutt, M. S. R. & Burkitt, D. P. 1965. Geographical distribution of cancer in East Africa: A new clinicopathological approach. In *Brit. Med. J.*, **2**:719–22

Huxley, J. 1942. *Evolution: The Modern Synthesis*, ed. John Randal Baker, reprinted 1974, London, Allen & Unwin

Iacono, J. M., Puska, P., Dougherty, R. M. *et al*. 1983. Effect of dietary fat on blood pressure in a rural Finnish population. In *American Journal of Clinical Nutrition*, **38**:860–9

Illmensee, K. & Stevens, L. C. 1979. Teratomas and chimeras. In *Sci. Amer.*, **240**:121–32
Imrie, J. & Imrie, K. P. 1979. *Ice Ages: Solving the Mystery*, London, Macmillan
IUCN Red List of Threatened Animals. 1988. IUCN, Gland, Switzerland

Jeon, K. W. & Jeon, M. S. 1976. Endosymbiosis in amoebae: recently established endosymbionts have become required cytoplasmic components. In *J. Cell Physiol.*, **89**:337–44
Johnson, A. D. & Herskowitz, I. 1985. A repressor (Mat 2 product) and its operator control expression of a set of cell type specific genes in yeast. In *Cell*, **42**:237–47
Jones, L. W. & Myers, J. 1963. A common link between photosynthesis and respiration in a blue green alga. In *Nature*, **199**:670–2
Jones, P. A. 1985. Altering gene expression with 5-Azacytidine. In *Cell*, **40**:485–6
Jukes, T. H. 1980. Silent nucleotide substitution and the molecular evolutionary clock. In *Science*, **210**:973

Kagawa, Y., Nishizawa, M., Suzuki, M. *et al.*, 1982. Eicosapolyenoic fatty acids of serum lipids of Japanese islanders with low incidence of cardiovascular disease. In *J. Nutr. Sci. Vitamol.* **28**:441
Kammerer, P. 1912. Adaptation and inheritance in the light of modern experimental investigation. In *Smithsonian Inst. Ann. Rept.*, Washington DC 421–41
Kammerer, P. 1923. Breeding experiments on the inheritance of acquired characteristics. In *Nature* **111**:637–40
Kammerer, P. 1924. *The Inheritance of Acquired Characteristics*, New York, Boni and Liveright
Kannell, W. B. and Dawber, T. R. 1972. Atherosclerosis as a pediatric problem. In *J. Ped.*, **80**:544–50
Karnovsky, M. L., Leaf, A. & Bolis L. C. 1988. Biological membranes: aberrations in membrane structure and function. In *Progr. Clin. Biol. Res.* New York, Alan R. Liss Inc
Kibball, A. P. and Oro, J. (eds) 1971. *Prebiotic and Biochemical Evolution*, New York, Elsevier
Keys, A., 1980. *Seven Countries: A Multivariate Analysis of Death and Coronary Heart Disease*, Cambridge MA, Harvard University Press
Koestler, A. 1967. *The Ghost in the Machine*, London, Hutchinson
Koestler, A. 1971. *The Case of the Midwife Toad*, London, Hutchinson
Koestler, A. 1978. *Janus: A Summing Up*, London, Hutchinson
Koestler, A. and Smythies, J. R. 1969. *Beyond Reductionism: New Perspectives in the Life Sciences*, Boston, Beacon Press
Kornitzer, M., de Backer, G., Dramaix, M. *et al.* 1983. Belgian Heart Disease Prevention Project: incidence and mortality results. In *Lancet*, **i**:1066–70
Kratochwil, K. 1983. Embryonic induction. In *Cell Interactions and*

Development: Molecular Mechanisms, ed. K. M. Yamada, New York, Wiley Interscience, 99–122

Kretsinger, R. H. 1977. Evolutionary considerations of calcium pumping by biological membranes. In *The Proceedings of a Joint US–USSR Conference*, ed. D. C. Tortenson, New York, Raven Press

Krinsky, N. I. 1966. The role of carotenoid pigments as protective agents against photosensitized oxidations in chloroplasts. In *Biochemistry of Chloroplasts*, ed. T. W. Goodwin, New York, Academic Press, **1**:423–30

Krinsky, N. I. & Margulis, L. 1968. Visible light: mutagen or killer? In *Science*, **160**:1256

Kuhn, T. S. 1970. *The Structure of Scientific Revolutions* (2nd ed), Chicago, University of Chicago Press

Kushner, D. J. 1971. Life in extreme environments. In *Chemical Evolution and the Origin of Life*, ed. R. Buvet & C. Ponnamperuma, Amsterdam, North-Holland, 485–91

Laga, E. M., Driscoll, S. G. & Munro, H. N. 1972. Comparisons of placentas from two socioeconomic groups. In *Morphometry Pediatrics*, **50**:24–32

Lamarck, J. P. 1873. *Philosophie Zoologique*, ed. Charles Martin (2 vols), Paris, Savy

Lamb, H. H. 1977. Climate past, present and future. In *Climatic History and the Future*, New York, Barnes and Noble

Lamb, M. J. 1977. *Biology of Ageing*, Glasgow, Blackie

Lands, W. E. M. 1986. *Fish and Human Health*, Orlando, Academic Press

Lange, R. 1966. Bacterial symbiosis with plants. In *Symbiosis*, ed. S. M. Henry, New York, Academic Press, **1**:99–170

Lanyi, J. K. 1980. Physical chemistry and evolution of salt tolerance in halobacteria. In *Limits of Life*, ed. C. Ponnamperuma & L. Margulis, Proc. Fourth College Park Symposium, Dordrecht, Holland, Reidel, 61–7

Lauer, R. M., Clarke, W. R. & Rames, L. K. 1978. Blood pressure and its significance in childhood. In *Post Grad. Med. J.* **54**:206–10

Laws, R. M. & Parker, I. S. C. 1968. Recent studies on elephant populations in East Africa. In *Comparative Nutrition of Wild Animals, Symp. Zool. Soc. London*, **21**:319–61

Ledger, H. P. 1968. Body composition as a basis for a comparative study of some East African mammals. In *Comparative Nutrition of Wild Animals, Symp. Zool. Soc. London*, **21**:289–310

Leyton, J., Drury, P. J. & Crawford, M. A. 1987. Differential oxidation of saturated and unsaturated fatty acids *in vivo* in the rat. *British Journal of Nutrition*, **57**:383–93

Leakey, R. E. 1981. *The Making of Mankind*, London, Hamish Hamilton

Leakey, R. E. 1982. *Human Origins*, London, Hamish Hamilton

Leakey, L. S. B. 1970. *Abbottempo Book 1*, London, Abbot Universal Ltd 2–7

Leakey, L. S. B. & Clarke, L. G. 1951. *The Miocene Hominidea of East Africa*, London, British Museum

Little, C. 1983. *Colonization of Land: Origins and Adaptations of Terrestrial Animals*, Cambridge, Cambridge University Press

Logan, R. L., Riemersma, R. A., Oliver, M. F. *et al.* 1978. Risk factors of ischaemic heart disease in normal men aged 40: Edinburgh-Stockhold study. *Lancet*, **i**:949–55

Lopez, A. & Crawford, M. A. 1967. Aflatoxin content of groundnuts sold for human consumption in Uganda. In *Lancet*, **i**:1351–4

Lowenstam, H. A. 1972. Phosphatic hard tissues of marine invertebrates: their nature and mechanical function, and some fossil implications. In *Chem. Geol.*, **9**:152–66

Low, M. G., Kincade, P. W. 1985. Phosphatidylinositol is the membrane-anchoring domain of the Thy-1 glycoprotein. In *Nature*, **318**:62–4

Lowenstam, H. A. & Margulis, L. 1980. Calcium regulation and the appearance of calcareous skeletons in the fossil record. In *Proc. Third Internatl. Symp. Mech*, Biomineralisation in vertebrates and plants, Kashikoyima, Mie, Japan, 289–300

Lynn, R. 1982. IQ in Japan and the United States shows a growing disparity. In *Nature*, 23 May 1982

McCarrison, R. 1953. *Nutrition and Health*, London, Faber & Faber/ McCarrison Society

McGinnis, W., Gerber, R. C., Wirz, J., Kuroiwa, A. & Gehring, W. J. 1984. A homologous protein coding sequence in *Drosophila* homeotic genes and its conservation in other metazoans. In *Cell*, **37**:403–8

Maclean, N. 1976. *Control of Gene Expression*, New York, Academic Press

Maclean, N. & Hall, B. K. 1987. *Cell Commitment and Differentiation*, Cambridge, Cambridge University Press

Maden, M. 1982. Vitamin A and pattern formation in the regenerating limb. In *Nature*, **295**:672–5

Maden, M. 1983. The effect of vitamin A on limb regeneration in *Rana temporaria.* In *Devel. Biol.*, **98**:409–16

Magnusson, M. 1980. *Vikings*, London, Book Club Associates

Malthus, R. 1798. *An Essay on the Principles of Population as it Affects the Future Improvement of Society*, ed. T. H. Hollingsworth, 1973

Mangelsdorf, P. C. 1964. Domestication of Corn In Darlington C.D. 1969, *Science*, **143**:538–45

Margulis, L. 1970. *Origin of Eukaryotic Cells*, New Haven, Yale University Press

Margulis, L. 1980. Flagella, cilia and undulipodia. In *BioSystems*, **12**:105–08

Margulis, L. 1981. *Symbiosis in Cell Evolution*, New York, Freeman

Margulis, L. & Schwartz, K. V. 1985. *The Five Kingdoms*, New York, Freeman
Marmot, M. G. and McDowell, M. E. 1986. Mortality decline and widening social inequalities. In *Lancet*, **ii**:274–6
Martin, P. S. 1966. Africa and Pleistocene Overkill. In *Nature*, **212**:339–42
Martin, P. S. & Klein, R. G. (eds) 1984. *Quarternary Extinctions, a Prehistoric Revolution*, Tucson, University of Arizona Press
Martin, R. D. 1983. Human brain evolution in an ecological context. 52nd James Arthur lecture on the evolution of the human brain. American Museum of Natural History, New York.
Marsh, O. C. 1896. *Dinocerata: A Monograph on an Extinct Order of Gigantic Mammals*. Monographs of the US Geological Survey, vol 10, Washington DC
Mason, B. 1963. Organic matter from space. In *Sc. Amer.* **208(3)**:43
Maynard-Smith, J. 1958. *The Theory of Evolution*, Harmondsworth, Penguin
Maynard-Smith, J. 1968. 'Haldane's dilemma' and the rate of evolution. In *Nature*, **219**:1114
McNamara, J. J., Molot, M. A., Stremple, J. F. & Cutting, R. T. 1971. Coronary artery disease in combat casualties in Vietnam, JAMA **216**:1185
Medawar, P. 1960. *The Future of Man*, London
Miller, J. R. 1983. 5s Ribosomal RNA Genes. In *Eukaryotic Genes: Structure, Activity and Regulation*, ed. N. Maclean, S. P. Gregory & R. A. Flavell, London, Butterworth, 225–38
Miller, S. & Stevenson, M. 1987. Food fortification: the need for scientific contribution. In *Biblthca.Nutr.Dieta*, **40**:82–95, Basel, Krager
Mingazzini, N. 1891. Sulla regenerazione nei Tunicata, *Bolletino Soc. Nat. Napoli, Ser.1*, year 5
Molecules of Life. 1986. Readings from the *Scientific American*, New York, W. H. Freeman & Co
Moncada, S. & Vane, J. R. (eds) 1983. Prostacyclin, Thromboxane and Leukotrienes. *British Med. Bull.*, **39**:1–300
Moorehead, A. *The White Nile*, London, Hamish Hamilton
Morgan, E. 1982. *The Aquatic Ape*, London, Souvenir Press
Morgan, E. 1986. Lucy's child. In *New Scientist*, **13**:5
Morgan, T. H. 1903. *Evolution and Adaptation*, New York, Macmillan (reprinted 1908)
Morgan, T. H. 1919. *The Physical Basis of Heredity*, Philadelphia, Lippincott
Morgan, T. H. 1910. Chance or purpose in the origin and evolution of adaptations. In *Science*, **31**:201–10
MR FIT Research Group. 1982. Multiple Risk Factor Intervention Trial: risk factor changes and mortality results. *JAMA*, **248**:1465–77
Muneoka, K. & Bryant, S. 1984. Regeneration and development of vertebrate appendages. In *The Structure, Development and Evolution of Reptiles*, ed. M. W. J. Ferguson, London, Academic Press, 177–96
Munroe, H., personal communication

Murray, G. W. 1951. The Egyptian climate in historical outline. In *Geographical Journal* **177**, London, Royal Geographical Society

Naegli, C. 1898. *A Mechanico-Physiological Theory of Organic Evolution*, trans. V. A. Clark & F. A. Waugh, Chicago, Open Court

Nagy, L. A. & Zumberge, J. E. 1976. Fossil micro-organisms from the approximately 2800 to 2500-million-year-old Bulawayan stromatolite: application of ultramicrochemical analyses. In *Proc. Natl. Acad. Sci. USA*, **73**:2973–6

Nass, M. M. K. 1969. Mitochondrial DNA: advances, problems and goals. In *Science*, **165**:25–35

NATO report, 1987 *Fat Production and Consumption – Technologies and Nutritional Implications*. Series A: *Life Sciences*, ed. C. Galli, E. Fedeli, New York, Plenum Press, **131**:131–44

Naughton, J. M., O'Dea, K. & Sinclair, A. J. 1986. Animal foods in traditional Australian aboriginal diets: polyunsaturated and low in fat. In *Lipids*, **21**:684–90

Newgreen, D. F. & Gibbins, I. 1982. Factors controlling the time of onset of the migration of neural crest cells in the fowl embryo. In *Cell & Tissue Res.*, **224**:145–60

Nishyama *et al*. 1962. Mexican forms of sweet potato. In *Econ. Bot.*, ed. C. D. Darlington 1969, **16**:305–14

Nordoy, A. 1987. Influence of (n–3) fatty acids on platelets and endothelial cells. In *Biblthca.Nutr.Dieta*, Basel, Krager, **40**:33–41

O'Dea, K. 1983. In *Agriculture and Human Nutrition*. Proceedings of Conference, Wononga, ed. K. A. Boundy, G. H. Smith, Aust. Inst. Ag. Sci., Dept of Agriculture, Vic. 56–61

Odell, G. M., Oster, G., Alberch, P. & Burnside, B. 1981. The mechanical basis of morphogenesis. 1. Epithelial folding and invagination. In *Devel. Biol.*, **85**:446–62

Okada, T. S. & Kondoh, H. (eds) 1986. *Transdifferentiation and Instability in Cell Commitment*, Osaka, Yamada Science Foundation

Olsen, E. 1970. The evolution of photosynthesis. In *Science*, **168**:438

Oparin, A. I. 1924. *Proskhozhozhdie zhiny*. Izd. Moskovski Rabochii, Moscow. (Oparin, A. I. 1924. *The Origin of Life*. Translated from Russian and reprinted in J. D. Bernal, 1967. *Origin of Life*, Cleveland, World, pp.199–241)

Oparin, A. I. 1938. *The Origin of Life*, from Margulis, S. New York, Macmillan

Oparin, A. I. 1969. *Biogenesis and Early Development of Life*, New York, Academic Press

Oro, J. 1962. Purine intermediates from hydrogen cyanide. In *Archs. Biochem. Biophys.*, **96**:293–313

Oro, J. & Kimball, A. P. 1961. Synthesis of purines under possible primitive earth conditions. Adenine from hydrogen cyanide. In *Archs. Biochem. Biophys.*, **94**:217–27

Oro, J., Holzer, G. & Lazcano-Araujo, A. 1980. The contribution of cometary volatiles to the primitive Earth. In *COSPAR Life Sciences and Space Research*, ed. R. Holmquist, New York, Pergamon Press, **18**:67–82

Osborne, G. 1967. Stages in the development of coronary disease observed from 1500 young subjects. In *Editions du Centre National de la Recherche Scientifique*, 15 Quai Anatole France, Paris 7e

Otterman, J. 1977. Anthropogenic impact on the albedo of the earth. In *Climatic Change 1*, 137–55

Packard, A. S. 1901. *Lamark: The Founder of Evolution. His Life and Work*, New York, Longman, Green

Pagels, H. 1983. *The Cosmic Code: Quantum Physics as the Language of Nature*, New York, Bantam Books

Painter, N. S. & Burkitt, D. P. 1971. Diverticular disease of the colon: a deficiency disease of western civilisation. In *Br. Med. J.*, **2**:450–4

Pellew, R. A. 1984. The feeding ecology of a selective browser, the giraffe (*Giraffa camelopardalis tippelskirchi*). In *Journal of Zoology*, London, **202**: 57–81

Personen, E. 1974. Coronary wall thickening in children. In *Atherosclerosis*, **20**:173

Pickett-Heaps, J. D. 1974. Evolution of mitosis and the eukaryotic condition. In *BioSystems*, **6**:37–48

Pilbeam, D. 1984. *Sunday Express Magazine*, 29 January 1984

Pirie, N. W. 1937. The meaningless terms of life and living. In *Perspectives in Biochemistry*, Cambridge, Cambridge University Press

Pirie, N. W. 1968. Organizer of discussion of anomalous aspects of biochemistry of possible significance in discussing the origins and distribution of life. In *Proc. Roy. Soc.* **B,171**:1

Pirie, N. W. 1972. Characteristics of living things. In *Biology of Nutrition, International Encyclopaedia of Food and Nutrition*, vol. 18., ed. R. N. T. W. Fiennes, Oxford, Pergamon Press

Pirie, N. W. 1987. *Leaf Protein and Its By-Products in Human and Animal Nutrition*, Cambridge, Cambridge University Press

Plate, R. 1964. *The Dinosaur Hunters: Othniel C. Marsh and Edward D. Cope*, New York, D. McKay

Plato, *Timaeus and Critias*, trans. Desmond Lee, 1971, London, Penguin Classics

Prescott, J. W., Nead, M. S. & Cairsin, D. B., 1975. *Brain Function and Malnutrition*, New York, John Wiley & Sons

Price, A. Weston. 1938. *Nutrition and Physical Degeneration*, La Mesa CA, The Price Pottingers Nutrition Foundation Inc

Puska, P., Salonen, J. T., Nissinen, A. *et al*. 1983. Changes in risk factors for coronary heart disease during 10 years of a community intervention programme (North Karelia Project). In *Br. Med. J.*, **287**:1840–4

Rambler, M. & Margulis, L. 1980. Bacterial resistance to UV irradiation under anaerobiosis: implications for pre-Phanerozoic evolution. In *Science*, **210**:638–40
Rattray Taylor, G. 1983. *The Great Evolution Mystery*, London, Secker & Warburg; New York, Harper & Row
Ridgway, S. H. (ed) 1972. *Mammals of the Sea: Biology and Medicine*, Springfield, Illinois, Charles C. Thomas
Ris, H. 1975. Primitive mitotic systems. In *BioSystems*, **7**:298–304
Rivers, J. P. W., Hassam, A. G., Crawford, M. A. & Brambell, M. R. 1976. The inability of the lion, *Panthera leo L.*, to desaturate linoleic acid. In *FEBS Lett.*, **67**:269–70
Rivers, J. P. W., Sinclair, A. J. & Crawford, M. A. 1975. Inability of the cat to desaturate essential fatty acids. In *Nature*, **285**:171–3
Rogers, J. 1976. Evolution of the eukaryotes. In *New Scientist*, **71**:33
Romer, A. S. 1954. *Man and the Vertebrates*, London, Pelican
Royal College of Physicians of London and British Cardiac Society. 1976. Prevention of Coronary Heart Disease. In *Journal of the Royal College of Physicians of London*, **10**

Sagan, C. 1980. *Cosmos*, New York, Carl Sagan Productions Inc
Sanders, T. A. B. 1983. Dietary fat and platelet function. In *Clin.Sci.*, **65**:343–50
Sarnthein, M. 1978. Sand deserts during glacial maximum and climatic optimum. In *Nature*, **272**:43–6
Sargent, T. D. 1969. Background selection of pale and melanic forms of the cryptic moth, *Philagalia Titea*. In *Nature*, **222**:585
Saxen, L. & Wartiovaara, J. 1984. Embryonic induction. In *Developmental Control in Animals and Plants* (2nd edn), ed. C. Graham & P. F. Wareing, Oxford, Blackwell, 176–90
Shah, K. P. & Shah, P. M. 1972. Relationship of weight during pregnancy and low birth weight. In *Ind. Ped.*, **9**:526–31
Shah, K. P. & Shah, P. M. 1975. Factors leading to severe malnutrition. A relationship of maternal nutrition and marasmus in the infants. In *Ind. Ped.*, **12**:64–7
Sharpe, E. J. 1970. *Ancient Empires*, London, Weidenfeld & Nicholson
Schimke, R. T. & Doyle, D. 1970. Control of enzyme levels in animal tissues. In *Ann. Rev. Bioche.*, **39**:929–76
Schmidt-Nielsen, K. 1964. *Desert Animals – Physiological Problems of Heat and Water*, Oxford, Oxford University Press
Schneider, S. H. & Londer, R. 1984. *The Co-evolution of Life*. Sierra Book Club
Schopf, J. W. 1973. On the development of metaphytes and metazoans. In *J. Paleontol.*, **47**:1
Schuster, W. & Brennicke, A. Genes travel out of cell nucleus! In *New Scientist* 26
Scolander, P. F., Hammel, H. T., LeMesurier, H., Hemmingsen, E. &

Garey, W. 1962. Circulatory adjustment in pearl divers. In *J. Appl. Physiol.*, **17**:184

Scott, P. P. 1968. The special features of nutrition of cats, with observations on wild felidae nutrition in the London Zoo. In *Comparative Nutrition of Wild Animals*, *Symp. Zool. Soc. London*, **21**:31–6

Shaper, A. G. & Jones, M. & Kyobe, J. 1961. Plasma lipids in an African tribe living on a diet of meat and milk. *Lancet*, **ii**:1324

Shaper, A. G. & Jones, K. W. 1962. Serum cholesterol in camel herding nomads. In *Lancet*, **ii**:1305

Shaper, A. G., Pocock, S. J., Walker, M., Phillips, A. M., Whitehead, T. P. & MacFarlane, P. W. 1985. Risk factors for ischaemic heart disease: the prospective phase of the British Regional Heart Study. In *J. Epidemiol. Community Health*, **39**:197–209

Sheets, H. & Morris, R. 1974. *Disaster In The Desert*, New York, Carnegie Foundation

Sheldon, P. R. 1987. Parallel gradualistic evolution of Ordovician trilobites. In *Nature*, **30**:561–3

Sheldrake, R. 1985. *A New Science of Life: The Hypothesis of Formative Causation*, London, Anthony Blond

Sidenbladh, E. 1983. *Water Babies*, trans. W. Croton, London, A. & C. Black

Sikes, S. K. 1968. Observations on the ecology of arterial disease in the African elephant (*Loxodonta Africana*) in Kenya and Uganda. In *Comparative Nutrition of Wild Animals*, *Symp. Zool. Soc. London*, **21**:251–73

Sinclair, A. J. 1975. Incorporation of radioactive polyunsaturated fatty acids in liver and brain of developing rat. In *Lipids*, **10**:175–184

Sinclair, A. J. & Crawford, M. A. 1973. The effect of a low fat maternal diet on neonatal rats. In *Br. J. Nutr.*, **29**:127–37

Sinclair, A. J., O'Dea, K. & Naughton, J. M. 1983. Elevated levels of arachidonic acid in fish from northern Australian obastal waters. In *Lipids*, **18**:877–81

Sinclair, H. M. 1968. Nutrition and atherosclerosis. In *Comparative Nutrition of Wild Animals, Symp. Zool. Soc. London*, **21**: 275–88

Sinclair, H. M. 1980. *Advantages and disadvantages of an Eskimo diet*. In *Drugs Affecting Lipid Metabolism*, ed. R. Fumigalli, D. Kritchevsky, R. Paolotti, Amsterdam, Elsevier, North Holland Biomedical Press, 363–70

Smith, G. R. 1981. DNA supercoiling: another level of regulatory gene expression. In *Cell,* **24**:599–600

Smithels, R. W., Shepherd, S. & Schorah, S. J. 1980. Possible prevention of neural-tube defects by preconceptual vitamin supplementation. In *Lancet*, **i**:339

Snow, M. H. L. 1981. Growth and its control in early mammalian development. In *Brit. Med. Bull.*, **37**:221–6

Spencer, H. 1864. *First Principles of a New Philosophy*, New York, Appleton

Spencer, H. 1887. *The Factors of Organic Evolution*
Spencer, H. 1893. The inadequacy of natural selection. In *Contemporary Review* 1893a. **43**:153–66; 439–56
Spencer, H. 1893a. Professor Weismann's Theories. In *Contemporary Review*:743–60
Sprecher, H. 1981. Biochemistry of essential fatty acids. In *Prog. Lipid Res.*, **20**:13–22
Stanier, R., Adelberg, E. & Doudoroff, M. 1963. *The Microbial World*, Engelwood Cliffs, New Jersey, Prentice Hall
Stanley, S. M., 1979. *Macro-evolution: Pattern and Process*, San Francisco, Freeman
Stanley, S. M. 1973. An evolutionary theory for the sudden evolution of multicellular life in the late Pre-Cambrian. In *Proc. Nat. Acad. of Sciences USA*, **70**:486
Stebbins, G. L. & Ayala, F. J. 1981. Is a new evolutionary synthesis necessary? In *Science*, **213**:967–71
Stirton, R. A., 1959. *Time, Life and Man*, New York, John Wiley & Sons
Stubbs, C. D. & Smith, A. D. 1984. The modification of mammalian membrane polyunsaturated fatty acid composition in relation to membrane fluidity and function. In *Biochim. Biophys. Acta*, **779**:89–137
Summerbell, D. 1981. Evidence for regulation of growth, size, and pattern in the developing chick limb bud. In *J. Embryol. exp. Morph.*, **65**:129–50
Sun, G. Y., Sun, A. Y. 1974. Synaptosomal plasma membranes: acyl group composition of phosphoglycerides and (Na+ + K+) – ATPase activity during fatty deficiency. In *J. Neurochem.*, **22**:15–18
Surgeon General's Report on Nutrition and Health 1988. Superintendent of Documents, Government Printing Office, Washington, DC 20402–9325
Sutcliffe, A. J. 1985. *On the Track of the Ice Age Mammals*, London, British Museum of Natural History
Swank, R. L. & Grimsgaard, A. 1988. Multiple sclerosis: the lipid relationship. In *Am. j. Clin. Nutr.* **48**: 1387–93
Szent-Györgyi, A. 1977. *Synthesis One*

Taylor, C. R. 1969. Metabolism, respiratory changes and water balance of an antelope, the eland. In *Am. J. Physiol.*, **217**:317–20
Taylor, C. R. 1970. Dehydration and heat: effects on temperature regulations of East African ungulates. In *Am. J. Physiol.*, **219**:1136–40
Taylor, T. G. 1970. How an eggshell is made. In *Sc. Amer.*, **222(3)**:89
Thele, D. S., Forde, O. H., Try, K. & Lehmann, E. H. 1976. The Tromso Heart Study: methods and main results of the cross-sectional study. In *Acta Med. Scand.*, **200**:107–18
Thurston, C. E. 1958. Sodium and potassium content of 34 species of fish. In *J. Amer. Diet. Assoc.*, **34**:396
Tinoco, J., Williams, M. A., Hincenbergs, I. & Lyman, R. L. 1971.

Evidence for non-essentiality of linolenic acid in the diet of the rat. In *J. Nutr.* **101**:937

Time Magazine, The heat is on, 19 October 1987

Treus, V. & Krevchenko, D. 1968. Methods of rearing and economic utilisation of eland in the Askaniya-Nova Zoological Park. In *Comparative Nutrition of Wild Animals, Symp. Zool. Soc. Lond.*, **21**:395–406

Trinkaus, J. P. 1984. *Cells into Organs: The Forces that Shape the Embryo* (2nd edn), Englewood Cliffs, NJ, Prentice Hall Inc

Turner, J. 1984. Why we need evolution by jerks. In *New Scientist*:34–5

Turkington, R. W. 1971. Hormonal regulation of cell proliferation and differentation. In *Developmental Aspects of the Cell Cycle*, ed. I. L. Cameron, G. M. Padilla & A. M. Zimmerman, New York, Academic Press, 315–55

Ucko, P. J. & Dimbleby, G. B. (eds) 1969. *The Domestication and Exploitation of Plants and Animals*, London, Gerald Duckworth & Co

Valentine, J. W. 1978. The evolution of multicellular plants and animals. In *Sc. Amer.*, **239(3)**:104

Van Valen, L. 1971. The history and stability of atmospheric oxygen. In *Science*, **171**:439

Van Valen, L. 1976. Energy and evolution. In *Evol. Theory*, **1**:179–229

Vitale, J. & Broitman, S. 1981. Lipids and immune function. In *Cancer Res.*, **41**:3706–10

Voelker, D. R. 1988. Phosphatidylserine translocation in animal cells. In *Biological Membranes: Aberrations in Membrane Structure and Function*, ed. M. L. Karnovsky, A. Leaf & L. C. Bolis *Progr. Clin. Biol. Res.*, **22**:153–64

Von Euler, U. S. 1982. Prostaglandins, historical remarks. In *Progr. Lipid Res.* Ed. R. T. Holman, Oxford, Pergamon Press Ltd, **20**:xxxi–xxxv

Waddington, C. H. 1947. *Organisers and Genes*, Cambridge, Cambridge University Press

Waddington, C. H. 1975. *The Evolution of an Evolutionist*, Edinburgh, Edinburgh University Press

Wallace, A. R. 1860. On the zoological geography of the Malay archepelago. In *J.Linn.Soc.* **4**:172–84

Wallace, A. R. 1896. The problem of utility: are specific characters always or generally useful? In *J.Linn.Soc. (Zool.)*, **25**:481–96

Wallenburg, C. S., Stolts, L. A. M. & Janssens, J. 1973. The pathogenesis of placental infarction. In *Am. J. Obstet. Gynaecol.*, **116**:835–40

Watanabe, I., Kato, M., Aonuma, H., *et al.* 1987. Effect of dietary alpha-linolenate/linoleate balance on the lipid composition and electroretinographic responses in rats. In *Advan. Biosci.*, **62**:563–70

Watson, J. D., Tooze, J. & Kurtz, D. T. 1983. *Recombinant DNA – A Short Course*, New York, Freeman & Co

Watson, W. 1949. The surface flint implements of Cyrenaica. In *Man*, **49**:100–04

Wayne, T. F. & Killip, T. 1967. Simulated diving in man: comparison of facial stimuli and response in arrhythmia. In *J. Appl. Physiol.*, **22**:167

Weber, P. C. 1987. n–3 fatty acids and the eicosanoid system. In *Fat Production and Consumption – Technologies and Nutritional Implications*, ed. C. Galli & E. Fedeli, Series A: Life Sciences, New York, Plenum Press Ltd, **131**:123–30

Weinberg, S. 1983. *The First Three Minutes: A Modern View of the Origin of the Universe*, London, Fontana

Weismann, A. 1880–2. *Studies in the Theory of Descent*, trans. R. Meldola (2 vols). London, Sampson Low

Weismann, A. 1892. *Das Keimplasma. Eine Theorie der Vererbung*, Jena, Gustav Fisher

Weismann, A. 1893. The all-sufficiency of natural selection. In *Contemporary Review*, **64**:309–38; 596–610

Weismann, A. 1893. *The Germ Plasm: A Theory of Heredity*, trans. W. Newton Parker and Harriet Roufeldt, London, Scott

Weismann, A. 1894. The Effects of External Influences on Development, Romanes Lecture, London, Frowde

Weismann, A. 1902. *The Evolution Theory*, trans. J. Arthur Thomson and Margaret R. Thomson, London, Edward Arnold

Wessel, G. M. & McClay, D. R. 1985. Sequential expression of germ-layer specific molecules in the sea urchin embryo. In *Devel. Biol.*, **111**:451–63

Whitehead, M. 1987. *The Health Divide*, London, The Health Education Council, 9–12

Whitehead, R. G., Paul, A. A., Black, A. E. & Wiles, S. J. 1981. Recommended Dietary Amounts of Energy for Pregnancy and Lactation in the United Kingdom. In *Food Nutr. Bull., United Nations University (Suppl.5)*, 259–65

Williams, G. 1981. Dietary deficiencies of captive dolphins. In *Prog. Lipid Res.*, **20**:259–60

Williams, G. & Crawford, M. A. 1987. Comparison of the fatty acid component in structural lipids from dolphins, zebra and giraffe: possible evolutionary implications. In *J. Zool.*, **213**:673–84

Williams, A. F., Gagnon, J. 1982. Neuronal cell. Thy-1 glycoprotein homology with immunoglobulin. In *Science*, **216**:696–703

Willis, A. L., Hassam, A. G., Crawford, M. A., Stevens, P. A. & Denton, J. P. 1981. Relationships between prostaglandins, prostacyclins and EFA precursors in rabbits maintained on EFA-deficient diets. In *Prog. Lipid Res.*, **20**:161–7

Wilson, E. O. 1975. *Socio-biology: The New Synthesis*, Cambridge, Harvard University Press

Wilson, E. O. 1978. *On Human Nature*, Cambridge, Harvard University Press

Winick, M. 1983. Nutrition, intrauterine growth retardation and the placenta. In *Trophoblast Research*, **1**:7–14

Woodland, H. R. & Old, R. W. 1984. Gene expression in animal development. In *Developmental Control in Animals and Plants*, ed. C. F. Graham & P. F. Wareing, Oxford, Blackwells, 422–504

Wright, S. 1981. Evolution in Mendelian populations. In *Genetics*, **16**:97–159

Yamanaka, T. 1966. Evolution based on comparative biochemical studies of cytochrome c. In *Ann. Report of Biol. Works. Science Faculty of Osaka Univ.*, **14**:1–37

Yamamoto, N., Saitoh, M., Moriuchi, A. *et al*. 1987. Effect of dietary alpha-linolenate/linoleate balance on brain lipid compositions and learning ability in rats. In *J. Lipid Res.*, **28**:144–51

Yokuyama, K. 1988. 29th International Conference on the Biochemistry of Lipids, Tokyo

Yu, T. C. & Sinnhuber, R. O. 1979. Effect of dietary w3 and w6 fatty acids on growth and feed conversion efficiency of Coho salmon (*Oncorhynchus kisutch*). In *Aquaculture*, **16**:31–8

Yudkin, J. 1972. *Pure, White and Deadly*, London, Davis-Poynter

Yudkin, J. 1985. *The Penguin Encyclopaedia of Nutrition*, London, Penguin

Zotin, A. I. 1972. *Thermodynamic aspects of Developmental Biology*, Basel, Switzerland, S. Krager

Zsmenhof, S. & Eichorn, H. H., 1967. *Nature* **212**:456

Index

About the Authors

Michael Crawford is head of the Department of Nutritional Biochemistry at the Nuffield Institute of Comparative Medicine at the Institute of Zoology in London. He holds a Special Chair at the University of Nottingham. Professor Crawford has an international reputation in the field of nutrition and has published two books and many papers on the subject.

David Marsh is a writer and researcher with a background in agriculture and philosophy. He has a special interest in the history of evolution theory and degenerative diseases.